ROUTLEDGE LIBRARY EDITIONS: COMPARATIVE URBANIZATION

Volume 3

THE MANAGEMENT OF HUMAN SETTLEMENTS IN DEVELOPING COUNTRIES

THE MANAGEMENT OF HUMAN SETTLEMENTS IN DEVELOPING COUNTRIES

Case Studies in the Application of Microcomputers

TIMOTHY J. CARTWRIGHT

LONDON AND NEW YORK

First published in 1990 by Routledge

This edition first published in 2021
by Routledge
4 Park Square, Milton Park, Abingdon, Oxon OX14 4RN
605 Third Avenue, New York, NY 10017

Routledge is an imprint of the Taylor & Francis Group, an informa business

British Library Cataloguing in Publication Data
A catalogue record for this book is available from the British Library

ISBN: 978-0-367-75717-5 (Set)
ISBN: 978-1-00-317423-3 (Set) (ebk)
ISBN: 978-0-367-77136-2 (Volume 3) (hbk)
ISBN: 978-0-367-77196-6 (Volume 3) (pbk)
ISBN: 978-1-00-317016-7 (Volume 3) (ebk)

Publisher's Note
The publisher has gone to great lengths to ensure the quality of this reprint but points out that some imperfections in the original copies may be apparent.

Disclaimer
The publisher has made every effort to trace copyright holders and would welcome correspondence from those they have been unable to trace.

The Management of Human Settlements in Developing Countries

Case Studies in the Application of Microcomputers

Timothy J. Cartwright

London and New York

First published 1990 by Routledge
11 New Fetter Lane, London EC4P 4EE

Simultaneously published in the USA and Canada by Routledge
a division of Routledge, Chapman and Hall, Inc.
29 West 35th Street, New York, NY 10001

Reprinted 1991

Typeset by LaserScript Limited, Mitcham, Surrey
Printed in Great Britain by
Antony Rowe Ltd, Chippenham, Wiltshire

British Library Cataloguing in Publication Data

Cartwright, Timothy J.
The management of human settlements in developing countries : case studies in the application of microcomputers.
1. Developing countries. Human settlements. Social planning
I. Title
307'.12'091724
ISBN 0-415-03124-9

Library of Congress Cataloging in Publication Data

Cartwright, Timothy J.
The management of human settlements in developing countries : case studies in the application of microcomputers / Timothy J. Cartwright.
p. cm.
Bibliography: p.
ISBN 0-415-03124-9
1. City planning – Developing countries – Data processing – Case studies. I. Title.
HT169.5.C37 1989 88-22967
307.1'2'0285416–dc19 CIP

Contents

Tables

Figures

Foreword

Human settlements are highly complex systems that are constantly changing in response to individual and group decisions and the modification of existing social, economic and physical patterns affecting them from within and without. So arises the need for planners and managers of urban areas to have ready access to related data and information and for the ability to manipulate them.

The use of microcomputers makes it possible to meet these needs especially in developing countries in which extensive technological and financial investments are not possible. The use of microcomputers is appropriate in that they allow the timely flow of information through inexpensive but sophisticated data-management systems. Although computerization is no universal panacea, it can in many cases render urban planning and management more effective and efficient. The United Nations Centre for Human Settlements (Habitat) recognized the important role microcomputers might play in human settlement planning and management, and in 1981 began providing advisory assistance, in response to specific requests from governments, offering expert technical advice on data management and human settlements and, sometimes, training programmes as well.

This book is based on information provided and sources consulted in the course of advisory missions undertaken in 1985 and 1986 by Mr T.J. Cartwright on behalf of the United Nations Centre for Human Settlements (Habitat). Although it is based on activities undertaken through its auspices following requests from governments, the Centre does not necessarily share all the views expressed in the book.

Special acknowledgement is due to Pergamon Journals Ltd, Oxford, for permission to use material in T.J. Cartwright, "Information Systems for Planning in Developing Countries", *Habitat International*, Vol. 11.1 (1987), pp. 191–205. Thanks are also due to the mission counterparts in human settlements agencies in the twenty countries whose experiences form the substance of this book.

This book refers to various proprietary products, marks and symbols (of which due acknowledgement is hereby made); in no case should these references be construed as endorsement of a specific product by the United Nations Centre for Human Settlements (Habitat) or any national government.

Dr Arcot Ramachandran
Under-Secretary General
Executive Director,
United Nations Centre
for Human Settlements

Introduction

Management is not a precise science; it depends on judgement and skill just as much as it does on analysis and technique. The purpose of this book is to present a series of case studies showing how art and science interact in practice. The case studies all deal with the use of microcomputers for the management of human settlements in developing countries. Since science alone cannot provide a blueprint for management, this book is aimed at encouraging people to learn as much as they can from the experience of others.

The main theme of the book is that microcomputers are appropriate technology for human settlements management in developing countries. Although the technology itself is relatively "high tech", its accessibility — in terms of both the cost of acquiring it and the ease of training people to use it — makes it appropriate for even the poorest countries. This is not to say that computerization is a panacea for all the problems of human settlements. On the contrary, computerization often creates problems as well as solves them; and there are problems for which computers are quite inappropriate. Nevertheless, there are many cases in management where electronic data-processing can make a significant contribution to improving effectiveness and efficiency. In such cases, it is argued here, the development of microcomputers has put computerization within the reach of all and made it an appropriate technology for developing countries.

Compared to other kinds of computers, microcomputers do have limitations: there are some things that big computers can do which small ones cannot. However, the higher cost and less "user-friendly" nature of mainframe and even minicomputers makes them a lot less appropriate than microcomputers for use in developing countries. Moreover, there is something intrinsically attractive about microcomputers and the way they put computing power directly and immediately into the hands of users. Of course, there is still a need for training and expert advice; but the idea of microcomputers is to let people take responsibility for their own computing needs without having to rely on a specialist intermediary. Thus, the focus of this book is on microcomputers, with a view to showing how much can actually be done with appropriate technology.

Related to this is another important theme of the book, and this is that the most critical factor in implementing computerization is probably access to equipment. Computer skills (unlike those involved in painting or music, for example) have little intrinsic appeal. Learning the commands of a computer programme is similar to memorizing telephone numbers. If you use them every day, learning them makes a lot of sense; if not, you soon forget what you have learned. So while training is a key element in any computerization strategy, there is little point to any of it unless staff members are going to have the motivation and opportunity to use computers in their work on a regular basis. Thus, the biggest single factor in developing computer literacy is likely to be the number of available machines, rather than how powerful or sophisticated they are.

The cases presented in this book are taken from reports on missions by the Special Adviser on Data Management of the United Nations Centre for Human Settlements (UNCHS, or Habitat), which is headquartered in Nairobi, Kenya. These missions were provided by UNCHS in response to specific requests from national governments for expert technical advice on matters related to data management and human settlements. Missions normally lasted from one to two weeks and sometimes included training programmes as well. UNCHS has been providing such missions since 1981 — before IBM even got into the microcomputer business. The missions used for this book all took place in 1985 and 1986 and reflect conditions and technology at the time. However, the management issues and principles that they reflect are essentially timeless.

It should be noted at once that the cases based on these reports represent no more than "snapshots" in time. None of the cases is meant to be either a complete or current picture of the situation in any country or in any governmental agency. Rather, the cases depict a situation from a particular vantage point at a particular point in time. Many, if not all, of the situations described in the cases will by now have evolved beyond what is shown here. Thus, the cases are presented here not to criticize what people may or may not have done in the past but to provide lessons from their experiences for the future.

The cases are organized into five sections, dealing with the following themes:

1. Getting started

Four cases dealing with first steps towards computerization, including how to get the most out of a computer consultant, how fast and how far to proceed in relation to other ongoing activities, how to evaluate a plan for computerization, and what to look for when a computer system is delivered and installed.

2. Computer applications

Four cases describing and discussing some of the multitude of possible applications of microcomputers to human settlements management, with emphasis on use of "off-the-shelf" rather than custom-written software.

3. Information systems

Four cases illustrating some of the problems and pitfalls involved in creating computerized human settlements information systems, including selection of key data, provision for keeping the data up to date, and anticipation of user needs.

4. Institutional factors

Five cases dealing with the relationship between computerization and the organizational environment, including identifying user needs, setting priorities for computerization, increasing productivity, improving effectiveness, and strengthening institutional capacity.

5. Policy choices

Three cases examining some of the broader policy implications of computerization, with particular reference to the balance of power within and between organizations, the impact of computerization on centralization and decentralization, and the contribution of computerization to the development process.

Each of the five sections is organized in the same way. There is an introduction setting out the theme of the section and summarizing each of the cases. Then there are the cases themselves. Finally, at the end of the book, there is a bibliography for each section pertaining to both the substance of the cases and the technical issues which they reflect.

Thus, a quick overview of the contents and scope of the book can be obtained by reading the introductions to each of the five sections. The five introductions can also serve as a guide to determining which cases are likely to be pertinent to a particular situation or problem that the reader is interested in, so those cases can be examined in detail. In this way, it is hoped that the experiences of others can be of practical benefit to people involved in the management of human settlements.

Part one

Getting started

Part one

Getting started

There is probably no clearer indication of the impact of microcomputers on human settlements management than that the question of whether to computerize is now hardly ever raised. Today it is just a question of how and when. Moreover, this transformation has been remarkably rapid. As late as 1984, data-management missions by UNCHS (Habitat) staff members would typically encounter opposition in principle to the use of computers in connection with human settlements. By the end of 1986, attitudes had completely changed. Now, UNCHS data-management missions are overwhelmingly about tactics — how many machines are needed, what kind should they be, whether it is wise to buy low-cost "clones" or only "brand-name" equipment, what software is required, who is going to provide training and so on. The main problem for UNCHS missions is now likely to be one of restraining excessive enthusiasm and making sure the technology is meeting genuine user needs. The change in attitudes on such a global scale over such a short time has been dramatic. For better or for worse, human settlements management will never be the same again.

Attitudes towards computerization have changed in another way, too. It used to be said that you should never start on computerization until you were sure that the corresponding manual process was working satisfactorily. There was little point in institutionalizing a process that was not ideal, it was said. Or that it would only compound the problem to try to do two things at once — namely, improve the process and computerize it at the same time. Now, as the technological mystery surrounding computers is beginning to dissipate, attitudes are becoming flexible. Perhaps computerization will help highlight some of the weaknesses in the way we are doing things; perhaps there are preferred ways of doing things that are possible only with the help of computers; perhaps computerization will help improve the underlying process rather than get in the way of changing it. Thanks to the gradual demystification of computers, people are coming to the conclusion that, when it comes to getting started with computerization, there is no time like the present.

Notwithstanding the new enthusiasm for computerization and in spite of the tremendous advances in "user-friendliness" that the microcomputer represents in comparison with its predecessors, it is rash to underestimate the complexity of what is involved in computerization. In other words, computerization is not always an easy thing to do. For one thing, there is a considerable initial investment to be made, an investment as much of time as of money. In particular:

- Although ready-to-use software packages, such as spreadsheets and database management systems, make things a great deal easier than they used to be, custom applications of these powerful tools still need to be designed. This is not always something that non-specialists are willing or able to take on.
- Once these custom applications are in place, data need to be entered and checked before the application can be used for any operational purposes. In the case of large-scale databases and information systems, data entry is an important consideration, if it has to be done by hand.
- To make effective use of all this capacity, users need to be trained, retrained and supported on a continuous basis.

Computerization is complex also because of the difficulty of foreseeing all its consequences. Ideally, computers allow an organization to do its work more efficiently and more effectively than it could by hand; but computers also let organizations do quite new things, which they could not do before, either because they did not have the time and resources or because computers give them new capabilities. In either case, success means enhanced performance. But success also alters the relationship between an organization and its environment, whether that organization is a governmental agency or a private company. Governmental agencies win and lose jurisdiction and responsibility, just as private companies increase or decrease market share. Through its impact on performance, computerization can play a role in enhancing or inhibiting this process of organizational development. Thus, effective computerization depends on anticipating not just its technical but also its sociological dimensions.

The impact of microcomputers on society may turn out to be akin to that of the automobile. Certainly, the long-run implications of microcomputers are likely to be as significant as those of the automobile. It is sometimes said that Henry Ford created the modern city not by actually planning it, but because his cars made it possible to create suburbia all over the USA and other industrialized countries. Microcomputers and the decentralized access to information that they can provide may likewise alter the urbanization trends that have been so much a feature of the second half of the twentieth century.

Another useful analogy that can be drawn between microcomputers and cars is that learning to use the first is about as difficult as learning to drive the second. In both cases, you can be a perfectly happy and successful user without being an engineer (although a rudimentary knowledge of "what goes on under the hood" does not do any harm). In both cases, learning how to operate the machinery depends on two key factors: motivation and access. Motivation is important because beginners might find the new technology strange and even frightening. Of course, people are naturally curious to make use of new technologies, and peer-group encouragement is often sufficient motivation — as long as there are no countervailing pressures such as suspicion or hostility from a superior. On the other hand, positive steps (such as overt encouragement, time off to take courses or bonuses for reaching certain skill levels) will naturally speed progress. Access to equipment is important both for learning how to use it and, later on, for remembering what to do. There is little that is intrinsically useful or valuable in learning to drive a car or use a computer. You have to practise in order to develop the skill in the first place; then, you have to practise if you want to "stay sharp". So, if you do not have access to the appropriate equipment (a computer or a car, as the case may be), the learning process is not going to be very efficient.

There is another important parallel between computers and cars when it comes to deciding what kind of equipment to buy. With computers, as with cars, there is always a temptation to buy more quality and more power than you really need. Who among us would not enjoy driving a Mercedes Benz or a BMW? Yet, when logic prevails, most of us settle for ordinary Fords and Toyotas. In organizations, however, the instinct of self-restraint is less common ("It's not my money") and that of self-preservation more common ("When in doubt, over-specify") than in personal affairs. The result is that corporate computers, like corporate cars, are often too good in quality and too few in quantity. Especially in the beginning, when the most pressing need is to give access to as many people as possible so that they can become "literate", it is sad to see resources tied up in fancy, brand-name equipment that is hardly used to its full capacities. Instead, while most people are just ordinary learners and users, the sensible policy is to buy as many basic, no-frills machines as you can. Then, when "power-users" have emerged, you can buy them sophisticated machines. Start with high-powered machines, and you will probably have too few for many potential users to get a chance to learn to drive.

Finally, dealing with computer vendors requires just as much skill and sales resistance as dealing with car salesmen (although the market has not yet bequeathed us a computer version of that particular breed, the used-car salesman!). In both cases, the most important rule is: try to be an informed consumer. Of course, there is much that the ordinary person does not know about computer technology, as about automotive technology. Nevertheless, the wise consumer will concentrate on what he or she wants the equipment to do and not be "snowed" by the salesman's blizzard of technical jargon. The wise consumer also knows that choosing the ideal machine for all occasions is next to impossible; so, he or she concentrates on avoiding disappointments — does it feel right when I drive it? is it similar to what my friends are using? does the dealer seem helpful and friendly? are parts and service readily available (not just from the dealer but from other sources as well)? Questions like these, which have become second nature when we buy a car, are just as applicable when we buy a computer system.

There is at least one difference in the evolution of the private car and the personal computer; but that, too, is instructive. This difference lies in the chronological relationship between the personal machine and its heavy-duty counterpart. In the automotive case, the private car led eventually to the invention of buses and trucks for moving large loads over standard routes. In the computer case, the reverse sequence occurred: microcomputers developed after their big brothers, mainframes and minicomputers. The result is that microcomputers have had to overcome a "credibility gap" that cars never had to contend with. No one thinks of cars as "just toys"; no one imagines that car drivers would really rather be driving buses or trucks, if only they had the opportunity and skill; and no one pretends that buses and trucks could meet all present and future transportation needs. Yet, those are precisely the arguments that microcomputer users have had to overcome in establishing their place in the computing environment.

This section presents four cases, all of which deal with first steps along the road towards computerization. In the first case, the Town Planning Department in Jamaica is thinking about computerization and is about to receive a consultant to advise on how best to proceed. The case describes what the Department can do to prepare for the arrival of its consultant. The second chapter discusses the case of the Town and Country Planning Division in Trinidad and Tobago. The Division is committed to computerization in principle but wonders whether now is the best time to do it. The third example presents the case of a planning agency in Tunisia that has commissioned consultants to prepare a plan for computerization and now has to decide whether to accept their recommendations. The fourth and final case in this section describes the Department of Public Works in Abu Dhabi (one of the United Arab Emirates), which is about to take delivery of a new computer system from a local vendor. The case discusses some of the problems and opportunities that can arise at this critical point in the computerization process.

Case 1: Preparing for a consultant (Jamaica)

In Jamaica, the Town Planning Department (part of the Ministry of Finance and Planning) is about to receive a consultant to advise how to proceed with computerization. Three priority applications have already been identified: management of development control on an island-wide basis; statistical survey analysis; and general office applications (including word processing). The purpose of this case is to suggest how the Department could actively prepare for the arrival of its consultant, without either anticipating or foreclosing on any recommendations he or she might want to make.

When thinking about computerization, one of the most important things to recognize is that it is a two-edged sword. When you computerize a certain process, you turn a spotlight on it. You peer in detail at every step and how it occurs. Inevitably, this raises questions about why it occurs: why do we need this particular piece of information? why does that agency have to approve things at this stage? who is actually responsible for this particular step? and so on. Even if these questions do not arise during computerization, they might later on, for, by making a process more efficient, computers raise questions about the value and efficacy of the process itself. In other words, computers can help you to see the forest in spite of the trees!

Thus, one of the most important ways for the Town Planning Department to prepare for its consultant is mentally. The consultant is going to ask awkward questions about why things are done the way they are, and the Department should be prepared to deal with these questions constructively rather than defensively. To this end, the case discusses a set of guidelines to help make the basic design decisions involved in computerization: assessing needs, identifying users, choosing software, choosing hardware and providing training. Finally, the case concludes with three main recommendations:

1. Try to be an educated and inquisitive client for your consultant. Even if computer technology seems very complicated, make your consultant explain things until you understand them.
2. Try to advance on several fronts at once. Just because a consultant is coming does not mean that all other inquiries and activities should cease.
3. Do not buy beyond your needs. With technology advancing as rapidly as it is, there is little sense in buying capacity beyond the ability to make immediate use of it. The chances are that hardware will be better and cheaper next year than it is this year.

Case 2: Now or later? (Trinidad and Tobago)

In Trinidad and Tobago, the Town and Country Planning Division of the Ministry of Finance and Planning faces a challenge on two fronts. For one thing, the Division is responsible for physical planning and development control throughout the country. However, the statutory basis for these responsibilities needs modernizing in response to the enormous social and economic changes that have occurred since independence. At the same time, the Division is under pressure to computerize. The government of Trinidad and Tobago is strongly committed to computerization as a means of improving public administration generally and human settlements management in particular. Thus, the Department faces a dilemma: whether to postpone computerization while statutory reform is carried out or delay statutory reform in order to proceed with computerization.

The case presented here proposes avoiding this dilemma by proceeding with the two initiatives in tandem. Rather than choosing between statutory reform or computerization, the case examines the implications of undertaking both simultaneously. In this way, computerization may be able to contribute to statutory reform by providing more information about development and development control than was hitherto available. Similarly, the process of reform provides a good opportunity for modifying the detailed forms and procedures of development control as well as its broad principles. Thus, statutory and administrative reform can benefit from and facilitate each other, if they are undertaken simultaneously.

Case 3: Assessing the plan (Tunisia)

In Tunisia, the Commission générale du développement régional et de l'aménagement du territoire (COGEDRAT) is the chief governmental agency responsible for regional development and physical planning. In 1983, the government decided that these functions should be strengthened through computerization. Accordingly, a team of outside experts was commissioned to prepare a *Plan informatique*, including complete specification of the hardware and software needed to implement it. On the basis of the resulting *Plan*, tenders were called for the supply and installation of a suitable computer system, and six bids were received. The purpose of the case presented here is not to review the bids (whose assessment would require detailed technical knowledge) but just to look at the *Plan* and how it is to be implemented.

The *Plan informatique* turns out to be ambitious in scope but vague in terms of its intended uses and users. Since, in fact, the *Plan* has been prepared not for COGEDRAT but for its predecessor agency, the proposals are inevitably somewhat out of date, but the biggest difficulty in assessing the *Plan* lies in the fact that its emphasis is on the collection and integration of large quantities of data, with relatively little indication as to how the data are to be collected and entered into the computer or how and by whom they will be used once they get there. Another complicating and somewhat sobering feature of the *Plan* is that development of software for the system (all of it apparently to be custom-written) is expected to take more than five years. Also, there is little attention paid to the use of microcomputers in general and to opportunities for communicating with the information system from the planning regions in particular.

Inevitably, therefore, the case points towards a rethinking of at least some parts of the *Plan informatique* in the light of both the enlarged mandate of COGEDRAT and developments in computer hardware and software over the past four or five years. The case recommends a more incremental strategy of computerization — one that is capable of adapting to changes in computer needs as well as in computer technology. In particular, the case suggests experimenting with a series of pilot information projects, before COGEDRAT commits itself to a single, national computer system.

Case 4: Taking delivery (Abu Dhabi)

The Department of Public Works in Abu Dhabi (United Arab Emirates) is the governmental agency responsible for the construction and maintenance of public buildings and works. The Department had just chosen a computer system from among four bids, based on specifications prepared by outside advisers. The purpose of this case is to discuss some of the desirable precautions to be taken during the critical period of installation of the equipment and development of appropriate software.

The case reviews the characteristics of the computer system chosen by the Department in view of its stated needs, and concludes that the capacity and flexibility of the system are somewhat limited. Accordingly, it is suggested that the Department should plan to replace rather than upgrade its new system. There is a discussion of various aspects of the supply and installation of hardware and software in the computer system and there are recommendations for clarification of some aspects of the warranty, service contract and training commitments incorporated in the contract with the supplier. The case endorses use of off-the-shelf (rather than custom-written) software wherever possible. Finally, there is a recommendation for action to strengthen in-house computer staff capacities in order to support users in the Department.

Case 1

Preparing for a consultant (Jamaica)

The Town Planning Department in Jamaica is getting a consultant to advise on computerization in the Department. Rather than just sit back and wait for the consultant to arrive, the Department should start now to prepare for his or her visit. This does not mean trying to anticipate or even influence what the consultant might recommend, but it does mean doing some preliminary "homework", to learn about computers and to think about how they might be applied to the work of the Department.

Moreover, it is important to recognize that computerization raises not just technical questions about means but also substantive questions about ends. You cannot do much to change the process of planning without also affecting the results. To give a simple example, computerizing departmental records (whether they are records of development-control decisions or library books) inevitably raises questions about what kind of information is being stored and what purposes the information really serves. In other words, when the means become orderly and efficient, the result is to focus attention on the ends. As departmental staff members and the general public become less preoccupied with the struggle to "get the job done", they start to think more about why and how the job is being done in the first place.

Thus, the more successful the Department becomes at using microcomputers, the harsher the spotlight that will be turned on the processes themselves. Computers are intended primarily for solving problems, but computers also raise questions by drawing attention to bottlenecks, inconsistencies, inefficiencies, resource misallocations, etc. in the way things are being done. In other words, introducing a new technology into an organization has ramifications that go far beyond the purely technical. The more the Department can "thrash out" some of these issues before the consultant arrives, the better he or she will be able to serve the Department.

Everyone has his or her own approach to using computers, but there are at least five main areas where early decisions are required. These are:

1. assessing computer needs;
2. identifying computer users;
3. choosing computer software;
4. choosing computer hardware; and
5. providing user training.

1. Assessing needs

In general, the Department should begin by identifying one or two priority areas for computerization and get to work on those. There might be some who argue that it is "rational" to identify all the potential uses of microcomputers and to concentrate on the "optimum" uses first, but this is neither necessary nor (probably) feasible. Certainly, it is going to be very difficult to identify all potential uses, since new and expanded uses will inevitably occur to staff members once they get their hands on the equipment; so, it is unlikely that anyone can know in advance what are the "right" priorities. Besides, computerization is also a matter of motivation; so, it might be a good idea to let people start with whatever they seem keen on doing. Thus, if some sensible priorities have been identified, use them. After all, there is not much in a planning office that cannot benefit from some degree of computerization, even if it is only word processing, and most microcomputers are flexible enough to be easily put to other uses as well as (or, even, instead of) those initially foreseen, if priorities shift somewhat with the benefit of experience.

The important thing is to have a clear and visible demonstration of success as early as possible. Other applications can come later — as long as hardware decisions are made wisely, of course (see below). Once there are one or two key applications working smoothly and effectively, there will be no shortage of suggestions as to other possible applications.

It is probably wise to start by applying microcomputers to processes that are already running smoothly on a manual basis — and not to abandon the old procedures until it is clear that the new ones are working properly, even if that means running the two systems in parallel for a certain period. On the other hand, it can also be argued that if a manual system is not working smoothly and has to be revised anyway, it makes sense to introduce microcomputers at the same time. Then, revisions made necessary or possible by the microcomputer can be incorporated at the same time. Either way, there are two important things to remember:

- Because they cannot tolerate ambiguity or lack of rules, computers tend to exacerbate rather than resolve any problems there may be in the way things are done.

- Because of Murphy's famous Law — "If anything can possibly go wrong, it will!" — it is always wise to keep the existing manual system in place until it is clear that the new computerized one is fully operational.

2. Identifying users

The single most important point in identifying potential computer users is that no one will use a computer if he or she cannot get to it easily. This is particularly true in the beginning, when scepticism and mistrust are high and uses are still relatively simple. Staff will quickly revert to manual methods they are familiar with, as soon as they have an excuse to do so — because they have to walk too far, they cannot get time on the computer when they want it, they do not like the room where the computer is located, the programme is too difficult to use, the programme still has bugs in it or the programme does not really do what they want it to do anyway.

Experience in many organizations shows that one microcomputer or one terminal can serve about three people. Naturally, this depends a lot on the nature and frequency of use, but three users per workstation is a good "ballpark" figure. Similarly, "the office next door" is about as far as you can expect people to go to use a microcomputer (except perhaps for batch-processing types of uses). The problem is that staff members do not perceive the journey to the computer room as a single trip. They know from experience that they will have to go back and forth several times to get things they forgot to take with them, to answer the telephone, to interact with their colleagues, to deal with crises, etc. They also know that going too far from their own office or desk takes them away from the environment where they are comfortable and have learned to operate effectively.

Finally, microcomputers need looking after. While it is important to make access to them as easy as possible for everyone, it is equally important to assign responsibility for supervising and looking after each unit on a daily basis to a specific person.

3. Choosing software

Software is one of the critical elements in any computer system. Fortunately, with the advent of microcomputers, the problems associated with developing software for specific applications have been simplified. A new category of software has emerged between high-level programming languages (such as FORTRAN or COBOL) and their customized applications. These new programmes are written for generic functions (such as word processing or database management) and are designed to be sufficiently "user friendly" for the user to adapt them to his or her specific needs. Some of these programmes are extremely sophisticated and well designed; they may represent many man-years and millions of dollars of development costs.

As a general rule, off-the-shelf, user-friendly generic software packages are preferable to custom-written programmes. The former are usually more flexible and more foolproof than the latter. Moreover, it is estimated that three-quarters of the needs of a typical planning office can be met with three such generic packages — word processing, spreadsheet analysis and database management. There are even some "integrated" packages offering all three capabilities in one, but these have not yet been as successful as the stand-alone packages.

To choose specific packages of this kind, it is usually enough to determine what are the "standard choices" — the programmes that are in widespread use — and then just assess whether any special departmental requirements may not be met and how critical this is. In other words, there is little need to innovate with new software, seductive as some of its advertising might sound: it is usually wise to rely on widely used (and widely proved) software. With standard software, it is easy to get help when staff members get stuck, there are ample materials available for training, there is a good chance of being able to upgrade to new versions of the same software and (when they learn to use these packages) staff members gain useful and transportable skills.

4. Hardware

There are five main questions to be dealt with here, most of them interrelated. There is the question of the microprocessor that lies at the heart of the computer and its operating system, of the size of its memory or RAM, of the size and speed of its mass-storage capacity, of the type and performance of its printers and of other peripherals (for communicating with other computers, plotting maps, etc.).

As a general rule, when an organization is just beginning to computerize, the important thing is to get as many workstations as the budget will allow. At the present time, a basic workstation means a microcomputer with 512 or 640 Kb, two floppy-disk drives, a graphics monitor and a printer — all of which should not cost more than about US$1,500, exclusive of local duty and taxes but including freight costs. In concentrating on numbers rather than sophistication, you maximize access for potential users; you minimize the effect of equipment failure, by providing redundancy; and you maximize the chances of being able to get prompt, reliable and low-cost service, if failure does occur. Simple equipment is least likely to go wrong, and, if it does, the consequences are unlikely to be disastrous. (As Bill Lear used to say when asked why his aircraft did not have a lot of fancy equipment in them, "If you don't have one, it can't go wrong!") The more users and the more files that are dependent on a single piece of equipment, the more catastrophic are the consequences of its failure.

Similarly, when it comes to mass storage and other peripherals, be careful not to take it for granted that bigger and more expensive is always better. If there is a clear and pressing need for hard disks, high-speed plotters and multi-user networks, then by all means get them; but if there are no such needs or if they are still only dimly perceived, funds are best spent — at least at the outset — on providing the widest possible access to computers.

Finally, compatibility and expandability are both important factors in the choice of equipment, particularly when procurement is likely to be incremental and modular. However, as long as purchases are confined to equipment that is compatible with a well-established standard and of a type that can be upgraded (increased memory, storage, etc.) in the years ahead, these issues should not cause serious problems.

5. Training

Motivation is the key to learning. Unless and until it is evident that microcomputers are going to make jobs easier or more pleasant than they are now, staff members are not going to make a great deal of effort to learn how to use the new equipment. People have to feel there is something in it for them before they will make a serious effort to become proficient.

This means several things. It means giving clear demonstrations of job-related uses, it means providing convincing evidence that staff members are going to have access to the equipment and software they require and it means not being too strict about personal uses of the equipment (such as playing games, typing personal letters or keeping track of private bank accounts). It also means encouraging staff members to learn on their own time and to help one another by sharing what they learn. For some of the generic software referred to above, there are good "hands-on" tutorial packages available to run on the computer itself; in other cases, it might be necessary to provide time off or, even, financial subsidy to ensure that staff members get the necessary training.

6. Conclusion and recommendations

Be an educated client

Just as citizens sometimes complain that they are no match for governmental experts at public meetings, ordinary people often feel they cannot challenge computer specialists. This assumption does little to promote communication between the two sides and should be avoided. There is much that people can do to become educated clients without becoming experts in their own right — just as you can be an educated consumer without becoming a scientist or an engineer. For example, departmental staff members should not let themselves rely exclusively on advice from a single source. They can identify three or four sources of information in addition to their consultant — such as vendors, users in other governmental agencies, magazines, books, etc. — and use them as a test of the views and opinions of their consultant. Similarly, departmental staff members should insist that their consultant take time to set out the issues and justify his or her advice in clear and simple terms. An expert should be expert enough to be able to explain things in plain language. In the end, it is easier for a consultant to work with an educated client than with an unresponsive one.

Advance on several fronts at once

The plan is to receive the consultant first and then buy the equipment based on the advice given. While this approach has much to commend it in terms of helping make sure that a sensible choice is made, there is a risk in it. The risk is that the equipment might not arrive and be fully operational until after the consultant has left, or that it arrives so late in the consultancy period that there is little or no time to help with the design and installation of software, the provision of training, etc. For these reasons, it might be wise (if somewhat unorthodox) to get at least one microcomputer before the consultant arrives. This will allow development work to begin almost immediately; it will allow some training or, at least, demonstrations of (for example) what a computerized development-control system would actually be like; and it will help add a dimension of reality to the eventual procurement decision that is all too often missing with first-time buyers. This first microcomputer could be borrowed, rented or otherwise acquired; but (for the reasons discussed above) it must be accessible to users in the Department. (It is no good just arranging for staff members to use a microcomputer in another Department; access must be immediate and convenient.) Even if this first microcomputer has to be bought outright and even if it turns out to be different from the ones eventually chosen, the cost will be minimal compared to the benefits derived from its joint use by the consultant and departmental staff members. Besides, the first microcomputer will still have a value and (as long as the choice was fairly conservative) will probably be compatible with whatever other equipment is later bought.

Do not buy beyond your needs

In the old days, buying a computer meant making a single large investment decision. It was important to try to anticipate not just present needs but long-term future needs as well, and to buy accordingly. Nowadays, things have changed. For one thing, microcomputer equipment is often upgradeable; users do not have to "buy ahead" of their needs as much they used to. Indeed, with technology advancing so rapidly and costs continuing to fall, it is unwise to buy too far ahead. For another thing, if it is well supported by service and software, microcomputers can be handed down; when one user outgrows what he or she is using, there is always someone else who can put it to use.

When computerization is introduced, the critical factor is always access — access to hardware, access to software and access to training. Quite inexpensive microcomputer systems are now capable of providing so much in the way of data-processing capacity that they represent a quantum leap for many planning agencies. The benefits of bringing this capacity to all parts of the Department are likely to be much greater than providing limited access to sophisticated equipment. So, the initial investment should probably go into the quantity of equipment rather than the power and capacity of individual workstations.

In the old days, access to computers was not so important, because the people who would ever actually use the equipment were more or less limited to computer specialists. However, computer technology has been profoundly changed by the microcomputer revolution: now, anyone can learn to use a microcomputer — and probably most should. In order to make computer literacy in the Department a reality, the key factor is going to be accessibility.

Case 2

Now or later? (Trinidad and Tobago)

Human settlements planners in Trinidad and Tobago face two major changes. First, planning is coming to the end of an era, and the institutions and procedures put in place at the time of independence now need review and reassessment in the light of experience and changing needs. Secondly, the growth and development of a modern society in Trinidad and Tobago, coupled with world-wide developments in micro-electronics, calls for an assessment of new methods and techniques available to planners.

The current planning process is laid down by the Town and Country Planning Act, Chapter 35:01, of the Laws of Trinidad and Tobago. The Act was originally drafted as Ordinance No. 29 of 1960, which in turn replaced the Town and Regional Planning Ordinance of 1939, the Slum Clearance and Housing Ordinance of 1939 and the Restriction of Ribbon Development Act of 1942. The object of the current law is "to make provision for the orderly and progressive development of land in both urban and rural areas and to preserve and improve the amenities thereof; for the grant of permission to develop land and for other powers of control over the use of land; to confer additional powers in respect of the acquisition and development of land for planning; and for purposes connected with the matters aforesaid. . .". The law was originally assented to as an Ordinance in August 1960 and formally proclaimed as an Act of the new government in August 1969.

The basic planning process was drafted in the late 1950s by an experienced UK planner and is based on three main principles:

first, all development of land is made subject to development permission from public authorities (section 8.1);

secondly, a hierarchy of formal planning documents is provided for, including a General Development Plan for the whole country as well as various regional and urban development plans (sections 5.1 and 5.2); and

> thirdly, it is the function of the responsible Minister to grant development permission in accordance with these plans and such other laws and policies as may apply.

The planning process is implemented primarily through the Town and Country Planning Division of the Ministry of Finance and Planning. The Division is organized into two main sections — Development Planning and Development Control — plus various administrative and support units.

For the purposes of planning, the country is divided into nine regions (eight in Trinidad and one in Tobago). These regions do not correspond with either the boundaries of local authorities (nine County Councils and four Urban Governments) or the regions used by other Ministries (except Education). The Division has four regional offices (one in Tobago), but these are used for development control rather than development planning.

Up to now, development planning in the Division has been focused primarily on preparation of various planning documents. Under the original Planning Ordinance, the Minister was required to produce a general development plan within four years. For most of the 1970s, this task was deferred (with the approval of Parliament), while various local and regional plans were prepared. In the period 1974–9, for example, the Division completed a National Framework, five regional development plans, 13 technical research papers for the Tobago regional plan and 10 local area plans. According to the Division (in evidence submitted to the Crooks Commission in 1979), it was felt that,

> since development planning was a continuous process which must be flexible and responsive to change, a one-off master-plan approach would not be adequate . . . [and] a whole series of documents would better . . . deal with different aspects of development and also with particular problems of selected regions and local areas.

By the end of the 1970s, however, the Division was advised that the legal basis for enforcement of development control without a statutory national plan was problematical. So, in 1978, the Division produced a draft National Physical Development Plan, which was adopted by Parliament in 1984. The purpose of the National Physical Development Plan is:

1. To make provision for the orderly and progressive development of land in both urban and rural areas;
2. To preserve and improve the amenities of urban and rural land; [and]

3. . . . to secure consistency and continuity in the framing and execution of a comprehensive policy with respect to the use and development of all land.

Thus, the Plan is "conceived as an instrument for securing rational, consistent land-use policies".

Development control by the Town and Country Planning Division is focused largely on processing applications for development permission and then monitoring for compliance. Under the laws of Trinidad and Tobago, any development requires the approval of at least two agencies before a building permit can be issued: Town and Country Planning Division and the Local Health Authority. In practice, the latter will not proceed until the approval of the former has been obtained.

Applications are received and, for the most part, processed in the Division's four regional offices. Staff members first review the application, including a site inspection wherever that is deemed necessary; then, at the discretion of the officer concerned, it may be sent to other governmental agencies for review; finally, a recommendation is prepared and forwarded to the Director of the Division for approval. Any case involving a change in land use affecting agricultural land automatically goes to a committee of the Cabinet. On approval, copies of the application are forwarded directly to the Local Health Authority (which might already have been consulted). If the application is refused, reasons must be given in writing. In any case involving erection of buildings or subdivision of land (but not change of land use), a preliminary application for "outline approval" may also be submitted. For this process, less information is required, and referral to other agencies is not normally undertaken. However, outline approval is normally subject to certain conditions and is in any case valid for only a year; moreover, referral to other agencies might be (and, in the case of the Local Health Authority, will be) still required before final planning permission can be obtained.

The Division has been processing about 10,000 applications a year for the last several years. About 15–20 per cent are outline applications; the rest are full-scale applications. Table 1 provides details of the number of applications received over the past ten years and how they were disposed of. As the data show, 70-80 per cent of the applications are approved; of the remainder, about a third are withdrawn by the applicant (sometimes for resubmission later on) or cancelled by the Division because they are inactive. More than 90 per cent of all applications involve residential uses; however, detailed data on the value of the land and/or improvements involved are not compiled. The Division is required by law to deal with any application within two months of receiving it. If an application takes longer, a schedule has to

Table 1 Applications for development permission and their disposition, 1975–84 (Trinidad and Tobago)

	New Applications Received	Held Over	*Disposition*				
			Approved	*Refused*	*Stopped*	*Total*	*Left Pending*
1975	6,798	49	5,257	1,335	183	6,775	72
1976	8,964	72	6,718	1,340	930	8,988	48
1977	10,487	48	7,579	1,506	1,315	10,400	135
1978	10,832	135	7,599	1,755	1,494	10,848	119
1979	11,896	119	8,080	2,328	1,452	11,860	155
1980	11,174	155	7,705	2,324	1,183	11,212	117
1981	10,708	117	7,624	1,827	1,195	10,646	179
1982	10,641	179	7,595	1,859	1,282	10,736	84
1983	10,926	84	7,302	2,384	1,252	10,938	72
1984	9,740	72	6,201	2,313	1,108	9,622	190
Annual average	10,222	109	7,166	1,897	1,139	10,203	117
Monthly average	852		597	158	95	850	
			Quarterly Breakdown, 1981-4				
1981 1st	2,670	117	1,882	492	294	2,668	119
2nd	2,757	119	2,010	421	346	2,777	99
3rd	2,749	99	1,889	477	297	2,663	185
4th	2,532	185	1,843	437	258	2,538	179
1982 1st	2,540	179	1,808	399	298	2,505	214
2nd	2,713	214	1,953	456	355	2,764	163
3rd	2,737	163	1,935	521	350	2,806	94
4th	2,651	94	1,899	483	279	2,661	84
1983 1st	2,972	84	1,995	606	310	2,911	145
2nd	2,897	145	1,903	656	365	2,924	118
3rd	2,782	118	1,874	636	301	2,811	89
4th	2,275	89	1,530	486	276	2,292	72
1984 1st	2,693	72	1,749	577	368	2,694	71
2nd	2,620	71	1,625	646	320	2,591	100
3rd	2,336	100	1,494	562	221	2,277	159
4th	2,091	159	1,333	528	199	2,060	190
Quarterly average	2,626	126	1,795	524	302	2,621	130

Source: calculated from data provided by the Trinidad and Tobago Town and Country Planning Division in May 1985.

be worked out with the applicant. In practice, the Division estimated for the Crooks Commission on Land and Building Use (1981) that 80-90 per cent of applications received are dealt with in the two months allowed. For the remaining cases (where extensions are required), total time taken for processing may amount to a year or more. There are several reasons for these delays, including time required for other agencies to complete their review and delays caused by the applicant's failure to provide information that is required. Taking all this into account, it is estimated that there are about eight or nine hundred applications under active review by the Division at any given time.

By and large, therefore, planning in Trinidad and Tobago still clings to the pattern established about a quarter of a century ago. The institutions, the processes and the techniques are more or less those that were originally envisaged. (In some cases, apparently, even the design of the forms is unchanged since independence or even before!) Naturally enough, the whole system is coming under increasing pressure as times change.

The pressure on the present planning system has been manifest in a number of proposals for reform. Among the more prominent of these have been the following:

1. evaluation of the organization and manpower resources of the Town and Country Planning Division in the light of its responsibilities, particularly in the area of monitoring and enforcing development controls (Report of the Crooks Commission, 1981);
2. integration and updating of land records in the Town and Country Planning Division, the Valuation Division and the District Revenue Offices (Statskonsult Report on Land Registration, 1980);
3. reform of the land-registration system to provide a uniform (and probably compulsory) basis for registering land title, based on a unique identifier for each land parcel and an adequate cadastral description (Statskonsult Report on Land Registration, 1980); and
4. separation of development planning and development control, with the first to be centralized and integrated with other kinds of national planning and the second to be decentralized subject to certain clear institutional constraints (Report of the Demas Committee, 1984).

Whatever one thinks of these various proposals, they clearly reflect a widespread feeling that it is time to make a fundamental reassessment of the process of town and country planning in Trinidad and Tobago, with a view to ensuring that it is capable of meeting the needs of the next 25 years as effectively as it has met those of the last.

Just as Trinidad and Tobago is reaching its own particular crossroads in terms of town and country planning, so too human settlements planning on a global scale also faces a sweeping set of technological changes. In the past few years, developments in microelectronics have fuelled a truly dramatic change in the ways we can process information: the key to this revolution is the microcomputer. Microcomputers are not just big computers in small packages; rather, "micros" represent a new approach to computing — an approach that computer specialists sometimes call "non-traditional". Microcomputers represent a qualitative change in computer technology because they are meant to be used by ordinary people — just as typewriters are.

What makes this change so significant for developing countries such as Trinidad and Tobago is its economics. Not only has hardware become smaller and software more flexible and "user friendly" than they were in the 1960s and 1970s, but both have become cheaper. For example, just in the period since 1980 (when, incidentally, a major series of computer feasibility studies was carried out for the government by an international consulting firm) computer capacity that seemed quite expensive for an entire Ministry can now be put on the desk of one of its staff members for only a few thousand dollars. To give a concrete example, all the information pertaining to the 800-900 "active" development-control files maintained by the Town and Country Planning Division could be recorded on a single two-dollar, shirt pocket-size floppy disk — even without any attempt to optimize systems or procedures. Not only would this save storage space, but it would enormously facilitate retrieval, analysis and processing of the files. It would be hard to exaggerate the significance of this kind of development in terms of its potential impact on the use of information in public administration generally and human settlements planning in particular.

It is difficult for people who have not actually watched the growth and development of microcomputers over the past ten years to appreciate the nature and scope of the changes in computing that micros have brought about. In fact, these small, cheap, flexible, easy-to-use computers have revolutionized the role of computers in planning and management. Consider the following developments:

- Microcomputers have altered the system-design strategy. In the old days, buying a computer meant making a single large investment; so it was important to try to anticipate long-term future as well as present needs and to buy accordingly. Now you can take an incremental approach. You can buy some equipment now and get the rest later as you need it. You want to try to assess the opportunity of being able to upgrade in the future rather than to buy that capacity now. Equally, you want to be sure that what you get now can continue to have an independent life of its own, regardless of what you decide to do about upgrading it.

- Microcomputers have also altered the nature of the purchasing decision. In the old days, most purchasing involved choosing a single system from a single manufacturer. Now you can proceed in a modular fashion. Compatibility is still an important consideration, but it is not the dominant factor it once was, because there is more standardization among microcomputers than there was in pre-microcomputer period. In the case of printers, for example, you might choose to buy from a manufacturer different from that of the microcomputer. In the old days, you agonized over making the "right" decision because you would be stuck with it for years; nowadays, you can afford to take a more aggressive attitude and try to stay abreast of the latest technology. Of course, it still pays to buy wisely, but you have a lot of latitude. In a way, this makes purchasing more of a challenge than it used to be.
- Microcomputers and their flexibility mean that you can adapt and adjust your requirements as you go along. In the old days, you used to have to make several fundamental decisions before you even began; for example, you had to decide whether you wanted a "scientific" or a "business" computer. Nowadays, you can get a general-purpose microcomputer and add to or enhance it as the need arises, e.g. with a mathematics co-processor or a colour-graphics board.
- Microcomputers with their user-friendly software allow everyone to be his or her own "user". In the old days, we talked about "d.p. specialists" on the one hand and "clients" on the other — and about the difficulties of getting them to talk to each other. Nowadays, learning to operate a computer is no more difficult than learning to drive a car or use a typewriter: even ordinary people can do it — as long as they have the opportunity and the motive to do so.
- Microcomputers mean that you do not have to worry about maintenance and repairs as much as you did. In the old days, computers meant special rooms with temperature controls, air filters, false floors and ceilings, air-lock doors, elaborate security systems, special power supplies, and so on. Nowadays, most of these would be for the comfort and convenience of the user rather than the requirements of the equipment. If it is really hot and

humid, you may find your microcomputer needs air-conditioning; if power fluctuates wildly, you may want to protect equipment against voltage spikes. However, apart from that, you can take your microcomputer almost wherever you go (if you want to); you can lend it to someone in another office; or you can take it home to work at night. If it does break down, service and parts for a microcomputer are invariably easier and cheaper to find than they would be for big computers. Security, on the other hand, does pose a new problem; in the old days, it was difficult for someone to just walk off with a computer!

The effect of this information revolution is to bring town and country planning in Trinidad and Tobago to another crossroads. Just as the process of indigenous development has brought increasing pressure to bear on a planning process designed a quarter of a century ago, so too global technological advances have contrived to offer new capabilities for enhancing that process. In a way, there is a happy coincidence in the arrival of the planning challenge and the availability of a low-cost technological response.

Case 3

Assessing the plan (Tunisia)

The Tunisian Commission générale du développement régional et de l'aménagement du territoire (COGEDRAT) is a parastatal body under the Ministère du Plan. COGEDRAT was created in 1986 by merging the Commissariat général au développement régional (COGEDER) with the Direction-générale de l'aménagement du territoire (DAT).

COGEDRAT is committed to computerization by virtue of its organizational heritage. For its part, COGEDER already had four microcomputers (an Apple II, an IBM PC and two IBM PC XTs) when it came into COGEDRAT. These were used principally for word processing, various database management applications in connection with the hundreds of regional development projects launched under the Sixth Plan, and development of a national population forecasting model by *gouvernorat*, using a spreadsheet.

By contrast, DAT came into COGEDRAT with no equipment but with a detailed plan for full-scale computerization. This *Plan informatique 1984–88* was prepared for the DAT by the Centre national de l'informatique in April 1983 and anticipated a total expenditure of more than DT600,000 over five years. Two years later, in April 1985, specifications based on the plan were sent to several computer-equipment vendors, and six detailed tenders were received.

As a result, COGEDRAT is faced with two important decisions: how best to make use of both the computer equipment and the computerization plan inherited from its antecedents and how to apply the substantial funds that had previously been allocated for computerization of DAT to computerization of COGEDRAT. It would be hard to underestimate the potential importance of these decisions; clearly, they will help determine the activities and capabilities of COGEDRAT for years to come.

According to the *Plan informatique* prepared for DAT, its purpose is to create "an information system for physical planning [*aménagement du territoire*] . . . defined as a data bank around which various data management capabilities are to be developed. . . . The data bank for physical planning consists of the collection and storage of all information relevant to physical planning, regardless of its eventual use, and its provision to all users" (*Plan*, p. 5). This data bank, which is forecast to grow to about 120 Mb, will cover all the "economic activities and the natural resources pertaining to man and his way of life" (*ibid.*, p. 41). Consequently, the *Plan* recognizes that the data necessary for the data bank will have to come from dozens of sources in numerous organizations and that keeping them up to date (whether this is annual, monthly or according to some other period) will be a constant and vital process.

Four principal "data-management capabilities" for the data bank are foreseen (*ibid.*, pp. 9-23):

- decision support systems, such as planning, simulation, forecasting, classification and factor analysis;
- analytical techniques similar to the above;
- statistical techniques, including questionnaire analysis and other statistical compilations; and
- plan management systems, including subsystems for development, monitoring and graphic representation.

It is estimated that these programmes will amount to approximately 12 Mb (*ibid.*, p. 59) over and above the 120 Mb for the data bank itself.

In order to implement this information system, the *Plan* (pp. 64-74) proposes procurement of the following equipment:

- central processing unit with 200 Kb of main memory, rising to 450 Kb by the fifth year;
- hard disk system with a minimum capacity of 250 Mb, rising to 350 Mb by the fifth year, with appropriate tape back-up system;
- operating console plus four workstations, with an additional eight to be provided during the later years of the project;
- a 300-line-per-minute printer, with an additional eight printers by the fifth year;
- plotter, digitizer, graphics monitors and photocopy machine by the fifth year; and
- software, including operating system and appropriate utilities, database management system and compilers for FORTRAN, COBOL and BASIC.

In order to establish and maintain the proposed information system, the *Plan* proposes a staff of three (an engineer and two programmers) as well as a computer specialist. It is estimated that 229 separate computer programmes will be required, representing approximately 215 person-months of development time (*ibid.*, pp. 82 and 87). These estimates apply only to development of the statistical techniques and plan management systems, with programmes for the other two applications (decision support systems and analytical techniques) being developed "as and when required" (*ibid.*, p. 90).

Thus, development of the main priorities of the *Plan* are expected to occur in accordance with the following timetable (with the number of months shown for full installation):

Plan monitoring subsystem	19 months
Graphic representation subsystem	22 months
Data management subsystem	52 months
Statistical analysis subsystem	58 months
Plan development subsystem	64 months

Presumably, any time devoted to development of other subsystems for decision support and analytical techniques as and when required will delay the above timetable accordingly.

A second assumption in the *Plan* is its exclusive orientation towards the establishment and use of a data bank. Without detracting from the importance (even the prime importance) of this kind of capability for an organization such as COGEDRAT, consideration should also be given to other kinds of computer applications. Among these might be the following:

- Word processing, especially for form letters, documents or reports subject to frequent revision, and "boiler-plate" text.
- Administrative functions such as accounting, personnel management, inventory and procurement.
- Project management, using techniques such as critical path method or PERT (Program Evaluation and Review Technique).
- Graphic applications other than cartographic and statistical ones, such as surveying and computer-assisted design (CAD).
- Communication with COGEDRAT regional offices and other computer-equipped agencies in the public and private sectors.

Naturally, it is impossible to do all this at once. Nevertheless, in planning a computer system to meet user needs over five years or more, applications such as those just mentioned ought at least to be considered, if only to reject or delay those which do not enjoy a sufficiently high priority.

As for the data bank that lies at the centre of the information system, there is a certain lack of precision in the *Plan*. In principle, the scope of the data bank is broad enough: it will include "all information relevant to physical planning regardless of its ultimate use". In practice, however, it is impossible to include all information that *a priori* might be relevant to a given issue. In practice, choices have to be made — often taking into account exactly what the "ultimate use" is likely to be. Since the *Plan* chose not to identify any specific uses or users, there is no real basis for making precise either the structure or the contents of the proposed data bank. Indeed, it is not even clear from the *Plan* (cf., pp. 5 and 59) whether the data necessary for the two principal data management systems (the plan monitoring system and the statistical techniques system) are to be part of the central data bank or represent a second and separate data bank.

Whether the computer system eventually consists of one or several data banks, the *Plan* fails to give any details about the parameters of the two main data management systems. At no point does it discuss precisely what data are to be managed, how they are to be processed and what the required outputs are (either on screen or from the printers). It is not enough just to say that the computer can provide whatever kind of results might be required: this may be true but it is hardly the basis for designing an efficient information system.

In any case, the proposed data bank is very big — 120 Mb. Yet, the only justification for its size offered in the *Plan* is a table (p. 41) in which it is not even clear whether the figures refer to the volume of data *currently* in use for planning or the volume of data that it would be *desirable* to use in the future. If it is the latter (as presumably it is), then surely some kind of explanation is required — unless the *Plan* would have us believe that planning always improves in direct proportion to the quantity of information that goes into it!

Similarly, it should be noted that the design of the data bank reflects a potential conflict between the need for historical information and the need to have up-to-date information (*ibid.*, pp. 5 and 9). This conflict stems from the fact that new information can either replace or be added to old information. However attractive the latter option may seem, it raises the spectre of an information system that is not just 120 Mb in size but which grows by 120 Mb each year or so!

The size of the data bank also means that there are likely to be problems with data-entry. Although the question of electronic transfer of data already computerized is not even mentioned in the *Plan*, it appears that most of the proposed data will have to be entered manually. Based on an average speed of two characters per second (as postulated in the *Plan*, p. 61), data-entry will occur at the rate of approximately 6 Mb per terminal per year. Assuming that COGEDRAT will want to enter all the data within a fairly short time (say, a year), not least because most data will need updating after a year, it appears that nine or ten terminals will be required exclusively for data-entry. After that, a similar number of terminals may have to be dedicated to updating the data bank. Data-entry can always be contracted out to a commercial agency, but this may be an expensive solution. Of course, still other terminals will be required for development and use of the programmes supported by the data bank.

To summarize, the information system proposed in the *Plan* is based on a traditional view of information as a resource that should be centralized in a single computer accessible from remote terminals. The number of terminals proposed in the *Plan* represents a relatively limited access. Other options, such as decentralized processing or liberal access, do not seem to have even been considered. In principle, centralization offers advantages of security and standardization, but the disadvantages are its high cost, vulnerability to failure, dependence on lines of communication that may be quite extended in the case of the regional offices and lack of flexibility compared to decentralized systems. Thus, when it is a case of information that is confidential or subject to frequent (e.g., daily) updating, security and standardization become important. But in other cases, decentralized and flexible options should be considered.

Furthermore, the proposed information system depends entirely on a herculean effort to develop a custom software package. Apart from the three compilers (FORTRAN, COBOL and BASIC), the only software proposed for the system is a database management package. Apparently even statistical analysis (for which there are well-known packages such as SPSS and SAS) is to be developed on a customized basis. No doubt that is why the *Plan* anticipates the need for so many programmes (229), such a long development time (64 months) and so many programming resources (17 person-years).

As a result of this approach, COGEDRAT will bear all the development costs of the system. By contrast, using existing off-the-shelf packages allows development costs to be shared with other users, actual and potential. It also significantly reduces the time required for "debugging", provides access to software based on hundreds of person-years of development, offers the possibility of subsequent upgrades to the software and provides a product that is likely to be flexible.

Because of its historical context, the *Plan* could not take account of the equipment and experience already developed in the former COGEDER. Thus, it is useful to review briefly the capacity and achievements of COGEDER, with its three IBM microcomputers, compared to those in the minicomputer systems offered in response to the call for tenders based on the *Plan*:

- The central processing unit (CPU) in each of the microcomputers has (or could have) a main memory of 640 Kb which is more than what is proposed in the *Plan*, even for the fifth year. The speed of the CPU (which is 4.77 MHz, compared to speeds of 6 to 12.5 MHz in the tendered minicomputer systems) could be effectively increased in various ways, including addition of a mathematics co-processor, such as the Intel 8087.

- The operating system of the microcomputers (PC-DOS) is adequate for simple tasks but is not realistically capable of multi-tasking or multi-user applications to the extent that most minicomputer systems are.
- The built-in hard disks on the microcomputers have a capacity of only 10 Mb each; this is much less than the hundreds of megabytes provided in the tendered minicomputer systems. However, it is possible to add external drives to the microcomputers to provide more than the 140 Mb specified. Corresponding tape systems are available for microcomputers, and microcomputers also have floppy disk drives (with a capacity of 360 Kb or more), which provide an easy and inexpensive way of transferring data and programmes from one machine to another.
- Microcomputer monitors and printers are substantially similar to those used on minicomputer systems. In both cases, monitors are available with graphics and/or colour capabilities, and printers are available for high-speed or letter-quality output.
- Similarly, although microcomputers are not currently equipped with either plotter or digitizer, both could readily be added.
- As far as software is concerned, the three microcomputers already have more software than is proposed in the *Plan*. Specifically, the microcomputers currently have two database management systems, three word processing programmes and three spreadsheet programmes — eight packages compared to the one proposed in the *Plan*. Moreover, the software generally available for microcomputers is significantly more extensive and usually less costly than minicomputer software, and the former is often easier to learn and use than minicomputer software.

In short, it is clear that use of the existing computer equipment in particular and of microcomputers in general should at least form part of any computer plan for COGEDRAT. For what uses and to what extent microcomputers should supplement or even replace minicomputers remains to be seen, but it is clearly a question that should not be ignored. In particular, examination of the division of labour between microcomputers and minicomputers is essential (1) to ensure the most effective use of the funds available for computerization and (2) to open up the possibility of acquiring specialized computer systems (e.g., for cartography or computer-aided design).

In retrospect, it is striking to see how the task of developing a computer plan for DAT has been carried out. On the one hand, responsibility for identifying user needs was assigned to technicians in the Centre national de l'informatique. On the other hand, the highly technical task of choosing one particular configuration from among the six tenders has been left to users in COGEDRAT. For whatever reason, the distribution of roles between specialists and generalists is exactly the opposite of what might have been expected. The result is that COGEDRAT staff are faced with a decision defined in large part by technical parameters: CPU cycles in nano-seconds, CPU power in MIPS (millions of instructions per second), communication speeds in bauds and so on. Yet, the critical question is really quite simple: which tender is capable of meeting COGEDRAT information needs at least cost? Beside that, all other questions are relatively unimportant.

This said, it is clear that everything depends on the definition of needs, and that definition is not always easy, especially in an organization which has relatively little computer experience and which is perhaps not fully conscious of all the various ways in which it currently manages information by hand. For an organization to define its information needs, therefore, it must overcome this state of innocence and become more aware of all the dimensions of information management. Computers are not like coats that you can just buy and put on. Computers represent a new way of thinking and a new method of working — which staff members must learn to understand and exploit. Before the *Plan informatique*, there was another study of DAT computer needs which wisely pointed out (p. 51) that,

> . . . it cannot be emphasized enough that the introduction of computers to DAT cannot by itself resolve all problems, especially insofar as the organization of information and the development of physical plans is concerned.
>
> Only institutional self-analysis by DAT staff . . . will lead to answers to these kinds of problems and to the most effective use of future [computer] systems.

Thus, the main conclusions to be drawn from this experience are:

- that COGEDRAT should take advantage of its recent creation to rethink the computerization plan inherited from DAT and to do so in the light of its responsibilities both for physical planning and regional development; and
- that, rather than delegate the task entirely to outside experts, COGEDRAT staff should assume a major role in this rethinking, with periodic expert assistance as required.

The result will be a plan that is solidly based in COGEDRAT's own experience and needs.

Furthermore, COGEDRAT (unlike either of its predecessors) has both a first draft of a plan and staff with practical computer experience. Accordingly, and given that certain applications will naturally have a greater priority than others, what is proposed is that a series of pilot studies be carried out. For example, COGEDRAT staff could establish a simple microcomputer-based plan monitoring system for both physical plans and regional development plans. It may be necessary to add to existing microcomputer resources (e.g., a mathematics co-processor, a large hard-disk unit, a digitizer, a plotter, software, etc.) but all on a scale far below that envisaged in the draft *Plan*. Similarly, it may be necessary to hire a consultant to assist with development of the pilot project for, say, four to six months.

In addition to providing useful results more quickly than envisaged in the *Plan*, this kind of incremental strategy of computerization will permit COGEDRAT to make a realistic and empirical assessment of information needs. In this way, staff will soon begin to appreciate what kind of information is really essential and what kind is merely desirable. Moreover, once committed to a comprehensive computer strategy such as the one proposed in the *Plan*, COGEDRAT may find it difficult to modify, much less abandon, the path it has chosen. By contrast, an incremental strategy risks not much more than the current incremental step.

Although an incremental strategy may seem more laborious and less ambitious than a strategy of total commitment, the former may well be more prudent. For one thing, a phased approach will allow COGEDRAT staff to acquire an understanding of their information needs as well as an appreciation of how better access to information affects needs. Just as getting a car for the first time will affect your perception of transportation needs, getting a computer will affect your definition of information needs. In other words, the information needs of COGEDRAT are not immutable; on the contrary, they will evolve in response to its experience and its resources.

Case 4

Taking delivery (Abu Dhabi)

The Department of Public Works (DPW) of the Emirate of Abu Dhabi is made up of four Divisions, namely: Service and Communications, Government Building Project Management, Finance and Technical Service. The first is responsible for the construction of airports, bridges, roads, harbours and marine works and for reclamation. The second is responsible for the construction of governmental offices, hospitals, clinics, schools, etc. The third is responsible for finance and administration, including budgeting, personnel, payroll and stores. The fourth is responsible for maintenance of public buildings and infrastructure and operates laboratories for materials testing.

Specifications for the computer system (see Table 2) were prepared by the general contractor responsible for construction of the new departmental headquarters building in Abu Dhabi, assisted by staff of the National Computer Centre. Specifications appear to have been based primarily on estimates of data-storage requirements rather than the uses or performance desired from the system.

In August 1984, tenders were invited for the supply of computer hardware, software, training and service as per system specifications. Four bids were received, each built around a different hardware configuration:

- Alpha Micro AM-1092, based on a 16-bit MC68000 CPU;
- Data General Eclipse S/130, based on a proprietary 16-bit CPU;
- Stratus/32 CP Systems based on proprietary CPUs; and
- Industrial Micro Systems IMS 8000S, based on 8-bit Z-80 CPUs.

In January 1985, the IMS 8000S bid was selected. Thus the DPW computer system will consist of 16 IMS Ultima III terminals, each connected to an IMS USER-8 board (a Z-80 CPU with 64 Kb dynamic RAM) and supported by an IMS 8000S file-server. The IMS 8000S is to be equipped with two eight-inch Winchester hard disks, each of 85 Mb capacity, and one 100 Mb streamer tape backup unit. Although not originally specified, the supplier will also provide one soft-sectored, double-sided/double-density (DSDD), eight-inch, floppy-disk drive with a capacity of 1Mb. Eight Okidata ML-84 dot-matrix printers will also be provided for direct connection to eight of the 16 terminals.

Table 2 Summary of annual requirements for data-processing in the Department of Public Works, in megabytes of computer storage, by divisions (Abu Dhabi)

1.	Service and Communications Division		
	(a) General typing: 1,000 documents at an average of 10 Kb each (plus 50% contingency)	15	
	(b) Document processing (reports, contracts, etc.): 18 different forms per file for 2,000 files at an average of 600 bytes per form (approx)	25	40 Mb
2.	Technical Service Division		
	(a) Laboratory reports: 13 different forms per file for 1,200 files at an average of 300 bytes per file (approx)	5	
	(b) General typing	5	
	(c) Directory of suppliers	15	
	(d) Filing	10	35 Mb
3.	Finance Division		
	(a) General typing: 1,000 documents at an average of 1 Kb each	1	
	(b) Personnel records: 5,000 files at an average of 1 Kb each	5	
	(c) Budget	14	
	(d) Stores inventory	3	23 Mb
4.	Government Building Project Management Division		
	(a) General typing	n/a	
	(b) Directory of consultants (Tenders Section)	20	
	(c) Directory of consultants (Design Section)	10	30 Mb
Total			128 Mb
Contingency (20%)			25 Mb
Grand Total			153 Mb

Source: Taken from the contractor's Schedule of Work for the Department of Public Works; translated from Arabic by EFHA Consulting Engineers.

All terminals and printers will be equipped with firmware that provides English and Arabic characters, with automatic context-editing for Arabic (so each character can be produced with a single keystroke). Software will include:

(a) an Arabic-English, menu-driven operating system (TurboDOS) with a password-based security shell;
(b) an Arabic-English word processing programme with on-line help menus (Wordstar, including MailMerge and Spellstar, the last in English only);
(c) an Arabic-English relational database management programme which includes a report generator and four-function arithmetic (dBaseII);
(d) an Arabic-English "Calculator" designed to work with dBaseII in order "to provide the features of most modern spreadsheet programmes"; and
(e) an Arabic-English "complete general accounting package" (presumably including general ledger, accounts payable, accounts receivable, payroll and inventory programmes) customized to the needs of DPW but based on existing software (such as dBaseII or the CYMA accounting package).

The supplier has also undertaken to provide the Department with high-level language compilers with appropriate editors and debuggers for FORTRAN, COBOL, PL/I, PASCAL, C and BASIC.

All the hardware is covered by a two-year manufacturer's return-to-factory warranty. The printers are covered by a similar 90-day (one year for the printhead) manufacturer's warranty. In addition, the supplier has agreed to provide a full warranty (parts and labour) on all hardware for a period of six months from the date of installation.

The supplier has agreed to install all the software within six months of installing the hardware, including preparation of various databases and report forms agreed upon with DPW. At the same time, the supplier will provide training in the use of the system and the software supplied with the system for up to six trainees and to a standard such that they are capable of training other users. The supplier has also offered (at extra cost) annual contracts for hardware maintenance (parts and labour) and/or software maintenance and updating.

Inasmuch as the DPW computer system has already been purchased, no assessment is provided here of either the original specifications for the system or the choice of hardware and software to meet them. Instead this case looks ahead to the question of getting the most effective use out of the new computer system.

If the computer system is used as proposed and we assume an ordinary rate of growth in its use, it is highly likely that DPW will reach some of the limits of its system within a relatively short period of time (say, a year). Among these limits and their likely effects are the following:

Word size: The DPW computer system is based on an eight-bit word size (internal data path and I/O bus). This is the smallest size now in use; at least within the CPU itself, many of the most popular microcomputers use 16-bit words (e.g., the IBM PC) and even 32-bit words (e.g., the Apple Macintosh). Word size affects both the speed of computation and the amount of memory that can be addressed.

Clock speed: The CPU speed (that is, the speed at which the computer can actually do its work) is 4 MHz. For computationally intensive operations (such as "number-crunching" or sorting records), slow clock speed (plus the small word size) means noticeable delays in the operations of the computer. It is difficult to give exact details on this, but it is conceivable that the time required to sort large databases will have to be measured in hours, not minutes or seconds.

Memory size: Because of the eight-bit word size of the system, the maximum amount of memory (RAM) addressable from any terminal is 64 Kb. Some of this space is taken up by the applications programme in use (e.g., the word processing programme or the database management programme), so that only what is left is available for the file being worked on (i.e., the report being typed or the database being accessed). In the case of word processing, for example, this may mean a document of only about 10 pages can be in memory at one time. Thus, working with large files may mean frequent disk access.

Hard disks: With 170 Mb of on-line disk storage (about 10 Mb per terminal), disk space is not excessive, although it may not appear as a problem in the short term. However, disk access may become a problem quite quickly. No information appears to have been provided to DPW on how far system performance in general (and disk access in particular) on the IMS 8000 is affected by multiple use, but it is quite likely that the effects will be noticeable and irritating to users before long.

Floppy disks: Floppy disks provide the primary means for loading software and existing electronic data files into the computer system or for physically moving files from one computer system to another. However, the eight-inch floppy disk provided with the system is a declining standard in today's computer market, particularly at the low end. Thus, even though the operating system

(TurboDOS) appears to be flexible enough, the lack of applications programmes in the eight-inch floppy disk format (MFM or FM) will be a perpetual constraint on expanding the range of uses of the computer system.

Printer output: Since the system is to include only eight printers, only eight of the 16 terminals will be able to generate their own hardcopy output. On the other hand, it is a simple matter (assuming funds are available) to add printers to the other terminals if desired.

Number of users: The maximum number of terminals that can be connected to the system is 16; this means that no further work-stations can be added to the system beyond the number initially supplied.

In summary, this means that the DPW computer system will probably be operating close to the limits of its hardware capacity soon after it is installed. There is some scope for expanding the system (e.g., by adding disk drives or printers) and/or upgrading it (e.g., by converting the "host" CPU to a 16-bit Intel 8088 chip, which is an option available from the manufacturer). But none of these changes is likely to have an impact on the performance of the DPW system as a whole and (except perhaps for the addition of a few printers) is unlikely to prove cost-effective in the long run.

As far as software is concerned, the DPW computer system is to be supplied with a user-friendly operating system (TurboDOS) and four applications programmes for word processing, database management, spreadsheet calculation and accounting. These programmes should go a long way towards meeting the Department's immediate data and text processing needs. However, three additional points need to be considered:

First, the supplier's proposal to provide spreadsheet capabilities with an add-on to dBaseII is ingenious but may not prove adequate. It may be necessary to obtain a proper spreadsheet package (e.g., Lotus 1-2-3 or Supercalc).

Secondly, there are several areas beyond the four specified where DPW may want to make use of its computer system. Among these are:

- project management (CPM/PERT)
- statistical analysis
- graphics and mapping
- computer-aided design (CAD)
- construction management/job costing
- internal electronic mail ("e-mail")

- communication with other computer systems.

All these applications will require additional software beyond what is supplied with the system and, in some cases, may exceed the capabilities of the hardware as well.

Thirdly, the choice of an eight-bit system based on the Z-80 CPU limits the system to the use of software written for such machines. There is quite a lot of software available for these machines (although by no means all of it will be readily available to the IMS 8000 system). However, most commercial microcomputer software is now being written for 16-bit machines and the larger RAM memory (typically 256 up to 640 Kb) which they make possible.

The supplier has agreed to supply and install all these programmes to meet the needs of DPW as set out in the original specifications. However, it is important for DPW to recognize that installing applications programmes for specific uses places demands not just on the supplier but also on the client. A client must be able to describe his requirements accurately and precisely in order for a supplier to be able to tailor software to meet them. In the case of DPW, this is particularly true of programmes for database management and accounting, where DPW has a number of very specific requirements regarding forms and procedures. The process of computerizing these functions will inevitably raise questions about how and why things are done the way they are. If the installation of the software is to be satisfactory, it is essential that DPW be prepared to work closely with the supplier on the design of the management systems the Department wants to have computerized.

The supplier has also made a number of commitments respecting warranty and service under the sales contract and (as an extra-cost option) for an extended period thereafter. It is important for DPW and the supplier to be agreed on all details of this aspect of the contract. Among the possible areas for misunderstanding in such cases are the following:

- the exact date of commencement of all warranties, including supplier's and manufacturers' warranties;
- the exact nature of performance standards in respect of customized software and user training;
- who is responsible and who is to pay for maintaining a suitable supply of spare parts; and
- the nature and extent of the penalties to be attached to failure to meet any of the above commitments.

In this connection, it is worth noting that the most critical failure is likely to be one that involves a hard disk. For most of the rest of the system (terminals, printers, cabling, etc.), there is sufficient redundancy built into the computer system to minimize the consequences of a failure. If one of the two hard disks fails, however, the capacity of the DPW system is immediately cut in half. If the second disk fails, the entire system will be completely inoperative. Hard-disk repairs are normally carried out at a factory. This means that service will inevitably take weeks rather than hours or days. Thus, replacement of a failed hard disk (at least temporary replacement while the original unit is being serviced) is the only way to keep the system going. Of course, the information stored on a hard disk that fails will be inaccessible, even if the disk is replaced by a functioning unit, and may even be lost for good. This underlines the need to make regular and frequent back-ups of important data files — although this will be a tedious business with only one floppy disk drive available.

Installation of a computer system has a human as well as a technical side to it. In addition to acquiring hardware and software, an organization must also decide who is going to use it and how its use is to be organized (including training and development). In the case of DPW, several options are under consideration, but no firm decisions appear to have been made at the present time regarding organization and staffing for the computer system.

In the light of the foregoing assessment, several conclusions can be drawn regarding the future use of the DPW computer system. In general, the Department should probably not plan to expand or upgrade the present system, except possibly to add a few printers as and when required. Instead, DPW should plan to replace the present system as soon as its limitations begin to affect its usefulness — a point which may in fact come quite soon. It is worth noting that printers from the present system will probably be compatible with the next one.

As far as software is concerned, DPW should be particularly careful to check that the promised spreadsheet capability meets its needs, and if not, to insist on a stand-alone package instead of the promised dBaseII "overlay". Moreover, for all the customized "overlays", DPW should ensure it gets all the information and code necessary for it to have the option of (a) running the generic programme without any overlay at all and (b) modifying the overlays provided or creating new ones.

As far as other software is concerned, the Department should ask the supplier to provide additional information about off-the-shelf software for other applications. In some cases (such as project management or statistics), it may be quite simple and economical to add at least limited capabilities. In other cases (such as computer-aided design), this may not be true and, given the relatively short anticipated lifetime of the present system, adding these capabilities may not make much sense.

During installation, DPW should be sure to check the following:

That training provided by the supplier is not confined to teaching people how to use menus in applications programmes but includes training in all the features of the software provided (TurboDOS, Wordstar, dBase, etc.).

That the supplier provides such clarifications as may be appropriate regarding standards of performance in respect of undertakings on warranties, training and software support and regarding the supply of additional software (including the high-level language compilers and related utilities referred to above).

That the supplier provides a complete set of technical/service manuals for all hardware supplied plus at least 16 sets of user manuals describing the operation of the terminals, the printers, the operating system and the other software and languages provided; and six sets of the training materials used by the supplier for training the six trainees under the existing contract.

Finally, it should be noted that, during the critical period of installation, DPW will have no independent source of technical advice. It is strongly suggested, therefore, that DPW secure the services of a qualified person over the next three to six months to have the following responsibilities and duties:

- to advise the Department on matters connected with the installation of the computer system, including physical location of the terminals, establishment of appropriate system operations procedures, design of the security system, training schedule, etc.;
- to work closely with the supplier during the installation of the computer system, including hardware, software and training;
- to advise the Department on additional applications of the computer system to the work of the Department;
- to advise the Department on organization and staffing of a suitable DPW data-management unit and on development of a long-term plan for development of the computer system (including procurement of a next-generation system and its possible integration with an interdepartmental computer network); and
- to make contact with other related governmental agencies (such as the Ministry of Public Works and Housing), with a view to learning from their experiences, exchanging software, co-operating on joint training activities, etc.

In short, microcomputers have brought about a revolution in computerization. For the first time, computer power is within reach, financially and technically, of ordinary people. However, these are potentialities: the actual benefits of computerization do not come automatically out of the box along with the equipment; they have to be earned. So it is important for individuals and organizations who want to take advantage of the new technology to prepare for it. It is not necessary for everyone to become a computer expert, but it is important for people to learn enough to be no longer complete innocents when it comes to buying and using hardware and software. Similarly, not every organization needs a "computer department", but every organization should take steps to ensure that suppliers do not take advantage of its inexperience and that users get the support they need to make maximum use of their equipment.

Computers are technically little more complex than many other machines that we use in everyday life, such as automobiles, cameras, television sets, video recorders and photocopy machines. Psychologically, the important thing is to overcome the feeling of complete helplessness that modern technology can sometimes provoke and to recognize that, even though there may be much that we do not understand about the technology of these machines, we can still learn to be moderately intelligent consumers and moderately skilful users.

Part two

Computer applications

Part two

Computer applications

One of the hardest things for people to appreciate, whether they are computer neophytes or experts, is the enormous range of possible uses of microcomputers. Of course, this potential is inherent in all computers, big and small, but it is the more than 10 million microcomputers in use today which, with their low cost and ease of use, have really begun to make this potential into a reality.

The key to this revolution has been software. Ten or fifteen years ago, when you bought a computer system, your software choices were very limited. Apart from the operating system and its utilities (which might include a primitive text-editor), all you could typically get were high-level languages (such as FORTRAN, COBOL or BASIC) and a few so-called "applications packages", such as a database-management system, a statistical package or a project-management system. Word processing was not available on computers; you had to buy special, dedicated hardware for that. Graphic applications of any kind were virtually non-existent. The electronic spreadsheet had not even been invented. In short, you could not expect to find as much as one one-hundredth of the quantity of software available for a typical micro-computer today.

Furthermore, there was little you could do with even this limited quantity of software without the help of a "programmer" familiar with the particular language or applications package that you wanted to use. Ten or fifteen years ago, software was designed for computer specialists, not ordinary users. Software was intended to provide an environment for programmers to write programmes in. Software was never meant to be a "consumer product" that users could just take out of the package and run. User programmes had to be custom designed and custom written. Thus, even though your accounting needs might be virtually identical to mine, we would each get our programmers to write custom accounting packages for us in COBOL or RPG, never thinking to merge our requirements and share the cost of developing a single programme.

Finally, since there were few software "standards" among different computers — even among those produced by the same manufacturer, let alone between those produced by different manufacturers — most software had to be bought specifically for your machine and (usually) from the same supplier as the hardware. Because of the relatively low volume of sales entailed in so fragmented a market, software prices tended to be high relative to what they are now. For example, there are not many microcomputer programmes that cost more than $1,000; a decade or so ago, there was very little software that could be bought for less than $1,000.

In short, the microcomputer revolution has been fuelled by software. There is more (and more user-friendly) software available for microcomputers than for any large computers. There is off-the-shelf, ready-to-go microcomputer software for almost every conceivable kind of data processing. There are programmes to chart the stars, calculate the tides and cast horoscopes. There are programmes to diagnose physical diseases and calculate mechanical stresses. There are graphics programmes for architects, designers and engineers. There are management systems for lawyers, doctors, dentists, consultants, churches and clubs. There are accounting packages, statistical packages, project-management packages, mapping and surveying packages, building-maintenance and vehicle fleet-management packages and so on. There are modelling and simulation packages for everything from the stock market to water-supply systems to population-migration patterns. There are generic database management systems and spreadsheet programmes whose flexibility and adaptability are limited only by the imagination of the user. There are numerous word processing programmes, as well as programmes that teach you to type, correct your spelling and improve your writing style. If all this is not enough, if you really cannot find a ready-made programme to meet your needs, there are literally dozens of programming languages available on microcomputers in which you can write your own programmes. Indeed, it is sometimes said that Visicalc (the first spreadsheet programme) did more than anything else to sell Apple microcomputers.

Much of this microcomputer software is user friendly. That is, the software is designed specifically for "naive" or non-specialist users. For example, microcomputer programmes often make extensive use of "menus" to show the options which the user has at each point in the programme. Similarly, microcomputer programmes often make liberal use of screen graphics (colour, reverse video, moving bars, overlapping windows, etc.) to make screens easy to read and understand. Microcomputer software may also have on-line help facilities: press a certain key and the programme will provide a summary of the things you can do. In some cases, the help is even "context sensitive"; that is, the information provided is adjusted according to the point in the programme that the user has reached. Finally, for many microcomputer programmes, there are computer-based tutorials and demonstration programmes, all aimed at teaching non-specialists how to use them.

Thanks to the emergence of standards among microcomputers (notably, the so-called IBM PC/compatible standard), the hardware manufacturers' monopoly on software has been broken. Most microcomputer software is now manufactured by independent software companies. They have a potential market of millions of microcomputer users, compared to only thousands a few years ago. Consequently, software prices have dropped by a factor of ten and software is sold in supermarkets and by mail order. In fact, there is a lot of perfectly good microcomputer software that is available free of charge from companies promoting their other products (such as magazines or bulletin boards) or from authors who simply are proud of their work and want to share it with others.

The significance of the development of all this software would be hard to exaggerate. When you read a book, you may get, in the few hours it takes you to read it, the benefit of perhaps five or ten years of work by one or two people. When you use a sophisticated microcomputer programme such as Lotus 1-2-3 or dBaseIII, you benefit in a few minutes from perhaps hundreds of person-years of labour and millions of dollars of research and development. Although it is a somewhat crude yardstick, the "knowledge multiplier" effect underlying the use of computer software is truly staggering. Each time you use a spreadsheet programme, for example, it is like having a team of support staff working for you — a team, incidentally, that always works quickly and accurately. Moreover, the team is always at your service, whether it is to help with a sophisticated econometric model or merely to produce a graph of your monthly bank balance over the past year. Of course, there are people who regard the use of computers for what they like to call "non-traditional" applications (such as word processing or drawing graphs) as trivial, but as people experience the convenience and low cost of using computers for all kinds of data processing, this argument loses credibility. Microcomputers are indeed like cars: people will use them whenever they find it cost-effective to do so.

This section and the next one contain eight cases chosen to illustrate some of the possible uses of microcomputers for the management of human settlements. The four cases in this section are meant to illustrate the wide range of possible applications, whereas those in the next section focus on different approaches to one specific application (namely, information systems).

Case 5: An embarrassment of choices (Burma)

Burma provides a classic example of how, even in a relatively poor country, the opportunities for computerization in support of human settlements management are legion. The computer revolution has only just begun to be felt in Burma on a wide scale; yet, the possibilities are no less in Burma than in many other countries. Given the inevitable shortage of resources — hardware, software and trained users — the first task is to choose priorities.

The case presented here deals primarily with the Urban and Regional Planning Division of the Housing Department (which is part of the Ministry of Construction), and, to a lesser extent, with other Divisions in the Housing Department and with the Rangoon City Development Committee (which is the principal municipal authority for Rangoon). The following sets of priorities are proposed:

- For the Urban and Regional Planning Division: statistical projections, project site planning, a database of map information, a housing-loan fund simulation programme and development of a geographic information system (GIS) for Rangoon.
- For other Divisions of the Housing Department: computerization of the property cadastre, administration and collection of ground rents and management of public housing units.
- For the Rangoon City Development Committee: collection of property tax and other forms of municipal revenue, administration of licenses, management of road maintenance and administration of building inspection.

The case concludes by stressing the overwhelming need, in a country like Burma, to make the most effective use of computing resources. This can be achieved by concentrating on three things. First, provide as many basic, no-frills, "plain vanilla" computers as possible and get them into the hands of as many potential users as possible. It is better to get three or four people using computers which they may one day outgrow than it is to give one person more computing capacity than he or she can make use of. If necessary, fast machines and elaborate peripherals can be provided later on — although (frankly) speed and sophistication are rarely critical to human settlements computer applications. Besides, it is surprising what a determined user can do with even the most basic equipment, when that is all there is.

Incidentally, there was another computer project in Burma which, by contrast, provided one mainframe computer to the Central Statistical Office at a cost equivalent to that of approximately 4,000 "plain vanilla" microcomputers! Even though a microcomputer does not have the potential speed or capacity of a mainframe, imagine how great would have been the impact of distributing 4,000 microcomputers throughout the Burmese public service.

There are two other critical factors in making the most of scarce computer resources. One is the importance of providing adequate support for training and applications development. You cannot just supply computers and expect people to teach themselves what to do. Of course, some people will do just that and they should be given every encouragement, but most people will need help, at least in getting started. The other factor in making the most of computer resources is to concentrate on opportunities for cost-recovery inherent in applications such as revenue collection and the recovery of arrears. It is quite possible that the potential for cost-recovery may turn out to be sufficient to cover the entire capital cost of computerization in less than a year.

Case 6: Physical planning applications (Yemen Arab Republic)

In Yemen, the Ministry of Municipalities and Housing is the agency responsible for implementing governmental policy in respect of human settlements in general and housing in particular. Its activities cover the full range of physical planning and management functions, including land-use planning, infrastructure development, construction management and housing administration.

For some time, officials have been contemplating the use of computers in the Ministry. The purpose of this case is to suggest some key areas where computers could have an impact and what kind of equipment would be most appropriate. Seven specific applications are identified:

1. land-use and infrastructure planning;
2. services and facilities planning;
3. land management;
4. plot and loan administration;
5. project management;
6. administration and finance; and
7. research and policy analysis.

The case recommends immediate provision of basic microcomputer capacity for all these applications through a graduated programme of investment in hardware and software. Secondly, the case emphasises the importance of a visible commitment by senior management in order to ensure the motivation and commitment of other staff. Thirdly, the case underlines the need for adequate training and support during the early phases of computerization.

Case 7: Social planning applications (Malaysia)

In Malaysia, the Ministry of Housing and Local Government is the principal agency of the federal government responsible for national policy on human settlements. It is divided into four departments and eight divisions. A few of these departments and divisions have access to computers; so there are some seven computer projects already under way in the Ministry. In 1985, the Ministry proposed acquiring a minicomputer with eighteen terminals for M$375,000, but this was rejected by the Treasury. Now the Ministry is faced with the problem of how to meet the growing demand for computers from its staff, with little prospect of approval for large expenditures.

The conclusion presented here is that the Ministry's computer needs are many and varied. Thus, small, decentralized systems, with lots of different, off-the-shelf software packages, may well make more sense than a large, centralized system on which most applications would need custom programming. On the specific issue of a planning information system (discussed in detail in Part three below), the case underlines the need for regular updating of the data in such a system and for adequate analytical tools if users are to be drawn to it. Such systems should be designed from the ground up rather than from the top down. Finally, the case discusses (once again) the need for adequate training and development. In addition to conventional mechanisms (such as seminars, courses and programmes), it is also important to capitalize on latent staff enthusiasm, to encourage staff to learn from each other ("peer" training) and not to discourage the informal role of self-appointed "computer gurus".

The case concludes by recommending continued incremental purchasing of microcomputer equipment, rather than commitment to a single, large system. In addition, the case calls for strengthening of in-house support systems, a programme of incentives for staff to improve computer literacy and planned investment in systems development using outside expertise where it is required.

Case 8: Management applications (Thailand)

In Thailand, the Land Transport Department of the Ministry of Communications is responsible for licensing and regulating all public vehicles. Private vehicles are licensed and regulated by the police, but their records are to be transferred to the Land Transport Department as soon as a suitable computer system has been installed. The Department is divided into seven divisions and 72 district offices, one for each province (*changwat*) throughout the country.

Given its decentralized operations and the quantity of data it has to handle, the overwhelming need for computerization in the Department is to help manage its records on a systematic basis — including records of vehicles, drivers, accidents, operators' names and addresses, etc. Naturally, there are several other potential applications in research, planning and departmental administration; but the main need is clearly database management. Faced with these requirements, the Department has embarked on a planned process of computerization. A computer data-entry system has been acquired, and work is now proceeding on the transfer of records to magnetic tape. For the future, there are plans for two further stages of procurement. The first is to be a three-workstation, multi-user system costing about US$24,000, while the second is to be a $1.2 million minicomputer system.

The purpose of the case presented here is to review both the definition of the problem and the design of the solution as proposed above. Thus, the case begins with a brief systems analysis of data-processing requirements in the Department. From this, it is apparent that while some of the Department's databases are quite big, the kind of processing they require is really quite modest.

From a system-design perspective, this has a number of important implications. First, data-entry becomes as important a design consideration as data-processing. Secondly, since almost all data originates in the *changwat* offices rather than at headquarters, there should probably be a reassessment of current plans to confine or even just to introduce computers to headquarters rather than to put them in the field. Thirdly, electronic data-communication between headquarters and *changwat* offices will be a key factor in keeping the databases up to date. On the other hand, it is not necessary for such communication to be instantaneous; for most departmental purposes, it will not matter much if data communication takes a few days rather than a few minutes.

For these and other reasons, the case concludes by recommending that the Department take another look at its computer needs before committing itself to the proposed system design. While computers are, in principle, general-purpose machines, some are manifestly more appropriate for some purposes than others. You do not need to be a technical specialist to know that, for example, if you want hundreds of people throughout the country to be able to use a computerized database, you are going to need hundreds of terminals for them to use; that the sooner that data can be put into electronic form, the more efficient and accurate all subsequent processing becomes; or that, if it takes two years for 20 keyboard operators to enter 20 per cent of a database, data-entry will not be complete for another eight years, unless additional resources are made available. There is nothing worse than spending a lot of money on a computer system only to find that it cannot do what you wanted it to do or that it can do it only with great difficulty.

The only way to avoid this kind of situation is to make a careful assessment of the desired computer applications and then to design the computer system accordingly. The range of possible applications is very broad, and the variety of hardware and software available is sometimes bewildering. It is often prudent to seek expert advice, but experts often differ in the advice they give. In the end, therefore, there is no substitute for "intelligent consumerism". Make sure you know just what you want a computer system to do for you, and use your common sense in asking whether a proposed configuration will actually be able to meet those needs.

Case 5

An embarrassment of choices (Burma)

According to one estimate, there are currently operating in Burma three or four mainframe computers, several minicomputers and about 30 microcomputers. The newest and biggest computer is an IBM 4381 with 60 Mb of memory and 7.5 Gb of disk storage. The system was installed in May 1986 in the Central Statistical Office (CSO) of the Ministry of Planning and Finance as part of a United Nations project (BUR/83/004). This project, which is due to run until the end of 1989, has also supplied 18 IBM microcomputers, eight of them with hard disks (and four of these with colour). The project also provides for extensive training in COBOL programming and systems analysis in a sequence of courses (based on the NCC training system) which last from one to five months each; already some 200 trainees have completed various parts of the programme. Other large computer installations in Burma are reported to include a VAX 11/780, a Honeywell Bull machine and an old ICL system in the University Computer Centre (UCC) at the University of Rangoon.

Turning now to the role of computers in the management of human settlements, resources are still relatively limited. In the Ministry of Construction, the Construction Corporation has just installed a PRIME minicomputer and two NEC microcomputers, but the Housing Department has no computers at all, although some staff members have been able to make occasional use of computers in other agencies. Similarly, the Rangoon City Development Committee (RCDC), which is responsible for the day-to-day administration of Rangoon, neither has nor uses computers at all.

The range of possible computer applications to the management of human settlements is extensive. Indeed, there is now so much ready-made software for microcomputers that, if the equipment is available, the main limitation on what it can be used for is the imagination and skill of its users. In practical terms, however, it does not make much sense to try to embrace all applications at once. It seems likely that equipment and training will only gradually become available to those working in

Table 3 Urban employment forecasts for Rangoon City and Division by area, showing the effect of different forecasting techniques (Burma)

Employment forecasts computed for the year . . .		1994	
Base 1974	Total employment	483.1	(thousands)
Years: 1984	Total employment	593.5	(thousands)
	Average annual increase . .	11.0	(linear)
	Annual rate of growth	2.06%	(exponential)

	Actual Employment		*Annual Growth Rate*		*Projected Employment*			*Weighted Mean Projctn*
Division	*1974*	*1984*		*Linear*	*Expon*	*Share*	*Shift*	
	(thousands)				(thousands)			
				1.0	1.0	1.0	1.0	Weights
Central	202.9	257.9	2.40%	312.9	327.8	325.9	327.9	323.6
Inner Urban	126.1	154.0	2.00%	181.9	188.1	188.5	188.3	186.7
Outer Urban	115.1	133.4	1.48%	151.7	154.6	156.0	154.3	154.1
New Towns	39.0	48.2	2.12%	57.4	59.6	59.6	59.6	59.0
City	483.1	593.5	2.06%	703.9	730.1	730.1	730.1	723.5
Outer Townships	84.6	112.9	2.89%	141.2	150.7	148.5	150.3	147.7
Total	567.7	706.4	2.19%	845.1	880.8	878.6	880.4	871.2

Source: All data are taken from *Rangoon City Structure Plan*, Vol. 1, Table 3, page 30, prepared by the Housing Department (Ministry of Construction) with the assistance of UNDP/UNCHS Project BUR/80/005 (Rangoon: mimeo, June 1986). The four projection techniques are taken from Neil Sipe and Robert Hopkins, "Microcomputers and Economic Analysis: Spreadsheet Templates for Local Government" (University of Miami, 1984), chapter 4.

Notes:

1. A linear projection assumes each division grows by the same average annual amount as it did in the past.
2. An exponential projection assumes each division grows at the same annual rate of exponential growth as it did in the past.
3. A relative-share projection assumes each division gets the same relative share of total growth as it did in the past.
4. A relative-shift projection assumes each division shares in total growth according to a linear projection of its past share.
5. The official forecasts are based on the assumption that each division grows at the same annual rate of exponential growth as the entire city did in the past.

the field of human settlements. Data, too, are more likely to become available in some areas than in others. Thus, the relative "pay-off" from computerization in some areas will likely be higher than in others for some time to come. Indeed, in operational areas, where computers can be used for improving the collection of rents, fees and taxes, computerization should be possible on the basis of full cost-recovery. Thus, the purpose of this case is to identify and discuss some possible priorities for computer applications, first, within the Urban and Regional Planning (URP) Division in the Housing Department and, then, in other Divisions of the Housing Department and the Rangoon City Development Committee (RCDC).

Discussions with Division staff served to identify a number of important potential computer applications. Five such uses are described and discussed in the following paragraphs.

1. Statistical projections

One of the basic elements of planning is forecasting. Yet, the available techniques are often complicated and tedious, even with the help of an electronic calculator. Of course, forecasts are only as good as the data on which they are based, but, for any given set of data, computers can do a lot to reduce the tedium and improve the quality of forecasting.

Table 3 provides an illustration of how spreadsheet programmes can be used to forecast population growth in small areas. The model forecasts urban employment in Rangoon in 1994, based on data for 1974 and 1984 and using four different techniques, as well as a weighted average of all four.

The purpose of the model is not to prove that one technique is better than another or even that any of them is appropriate to this particular case. (The only way to determine that is to wait until 1994!) What the model does is to relieve the planner of the need to calculate each projection by hand, so that he or she can concentrate on assessing the assumptions and implications that lie behind each technique and how these apply to the specific point at issue (employment in Rangoon). In that way, the planner is encouraged to use his or her knowledge and experience to judge what, in this particular case, is likely to be the best approach to projecting the future. Anyone can do the arithmetic involved in these techniques: the planner's skill derives from knowing when and how to use them. Computers can help planners concentrate on this second issue by doing the first one for them.

2. Project design and preparation

One of the activities of the Division is the design and preparation

Table 4 Annual capitalization requirements for a hypothetical revolving fund for housing loans (Burma)

Housing Costs	Unit Cost in Year 1 (Kyats)	20,000
	Downpayment required (%)	30
	Estimated annual inflation in construction (%)	10
Loan Terms	Maturity of loan (years)	10
	Rate of interest (%)	5
Operating Costs	Administrative costs (% of loans outstanding)	0
	Estimated collection rate (% of repayments due)	100

Year	No. of Units	*Annual Disbursements*	*Total Loans Outstanding*	*Annual Payments*		*Reqd Annual Capitalization*
				Principal	*Interest*	
1	400	5,600,000	5,600,000	0	0	5,600,000
2	600	9,240,000	14,280,000	560,000	280,000	10,080,000
3	1,000	16,940,000	29,792,000	1,428,000	714,000	19,082,000
4	1,000	18,634,000	45,446,800	2,979,200	1,489,600	23,102,800
5	1,000	20,497,400	61,399,520	4,544,680	2,272,340	27,314,420
6	1,000	22,547,140	77,806,708	6,139,952	3,069,976	31,757,068
7	1,000	24,801,854	94,827,891	7,780,671	3,890,335	36,472,860
8	1,000	27,282,039	112,627,141	9,482,789	4,741,395	41,506,223
9	1,000	30,010,243	131,374,671	11,262,714	5,631,357	46,904,315
10	1,000	33,011,268	151,248,471	13,137,467	6,568,734	52,717,468
11	1,000	36,312,394	172,436,019	15,124,847	7,562,424	58,999,665
12	1,000	39,943,634	195,136,051	17,243,602	8,621,801	65,809,037
13	1,000	43,937,997	219,560,443	19,513,605	9,756,803	73,208,405
14	1,000	48,331,797	245,936,196	21,956,044	10,978,022	81,265,863
15	1,000	53,164,977	274,507,553	24,593,620	12,296,810	90,055,406
16	1,000	58,481,474	305,538,272	27,450,755	13,725,378	99,657,607
17	1,000	64,329,622	339,314,066	30,553,827	15,276,914	110,160,363
18	1,000	70,762,584	376,145,244	33,931,407	16,965,703	121,659,694
19	1,000	77,838,842	416,369,562	37,614,524	18,807,262	134,260,629
20	1,000	85,622,727	460,355,332	41,636,956	20,818,478	148,078,161

Annual Disbursements is the number of units times the unit cost less the downpayment adjusted for the rate of inflation of construction costs.
Total Loans Outstanding is the cumulative balance of Annual Disbursements less the Principal Repayments.
Principal Repayments is the Total Loans Outstanding in the previous year divided by the loan maturity period.
Interest Repayments is the Total Loans Outstanding in the previous year times the rate of interest on the loan.
Required Annual Capitalization is the difference between the Annual Disbursements and Repayments adjusted for administrative costs and collection rate.

of public-housing projects, including sites-and-services schemes. Each such project naturally requires extensive planning and costing prior to implementation. The essential part of this process is a comparison of layouts which must be compared from both a social and a financial point of view in order to try to provide the maximum level of amenities at the minimum unit cost. The procedure for making these comparisons can be made fairly routine. The problem is that each of the numerous possible layouts involves extensive recalculation of quantities and costs for every aspect of the project. Once again, the tedium of doing these calculations often discourages the planner from examining all the options — and from questioning the assumptions on which the calculations are based.

There are computer models which can do all the necessary calculations regarding plot sizes, construction costs, infrastructure costs, etc., and leave the planner free to concentrate on the planning issues. One such model designed for use on microcomputers is the famous "Bertaud Model" developed by and available from the World Bank.

3. Map database

Another fundamental aspect of planning is the use of maps. In most planning situations, the number of maps soon becomes large enough to require some sort of catalogue of coding system to keep track of them. The Urban and Regional Planning Division has about 500 maps on microfilm for which a catalogue system is urgently required.

The management of data records (database management) is a traditional area of computer application, not just for maps but for any kind of standardized records. In the case described here, the requirements are well within the capacities of a simple microcomputer using off-the-shelf software.

4. Housing-loan fund simulation

Division staff have been considering various schemes for financing the construction of new housing. Several different proposals have been discussed, one of which is to establish a revolving fund financed partly from annual loan repayments and partly by an infusion of new capital.

The viability of financial schemes such as these depends to a considerable extent on their operating parameters. If loans are increased too quickly or if repayments are not fast enough, if interest rates on the loans do not match those in the construction sector, if administrative costs are too high and collection rates too low — all of these factors affect how far the fund can be really self-supporting and how much it will have to depend on additional capitalization. Microcomputers can be readily used for financial models of this kind. Table 4 provides an example.

In this particular scenario — the parameters of which are laid out at the top of Table 4— with a loan rate of only half the estimated rate of inflation in the construction sector, it is clear that the fund cannot become entirely self-sustaining as long as new loans are being provided (although the annual capitalization required grows at a fixed rate from about the thirteenth year onwards).

5. Geographic information systems (GIS)

Much of planning deals with spatial data; so computers can do a lot to help planners by storing, processing and displaying information on a geographical basis. Such capabilities are called geographic information systems (or GIS).

For example, a typical GIS package can store, analyse and map data pertaining to points, areas and networks. This gives the planner the capability

- to find the perimeter or area of a region;
- to calculate averages and variances of variables;
- to analyse variables by regression or gravity models;
- to find points within a region or a circle;
- to find the optimum location in a plane or network; and
- to find the shortest path from one point to another,

as well as to draw maps illustrating any or all of the above results. Of course, the maps produced by a simple microcomputer-based GIS will not be of the same standard as those produced by proper cartographic equipment, but they are adequate for analysing and presenting the results of planning analyses or proposals.

For example, the maps in Figures 1 and 2 were produced by a typical microcomputer GIS package using data from the 1986 *Rangoon City Structure Plan*. The first map shows the average population density in each of the eight regions of Rangoon Division. The second map shows the highway and railway system in Rangoon Division, as well as the three lakes (Kanadawgi, Inya and Hlawi). Note that the lower map is an enlarged (or "zoomed") part of the upper map. Both maps were produced on a standard microcomputer and dot-matrix printer using the Urban Data Management Software (UDMS) package available without charge from the United Nations Centre for Human Settlements (Habitat) in Nairobi.

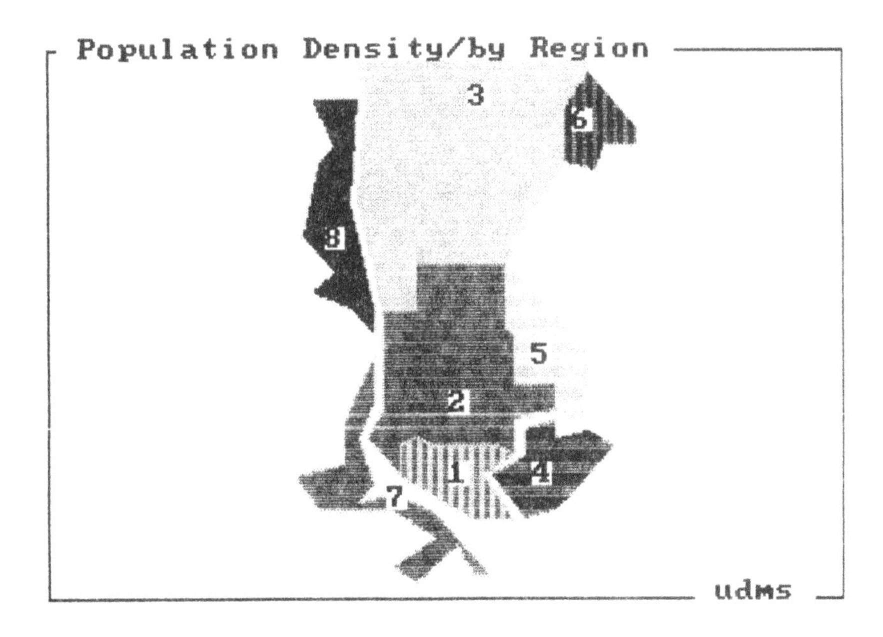

Variable Legend: Population Density

>= 1,750 < 4,222	>= 14,111 < 16,583
>= 4,222 < 6,694	>= 16,583 < 19,056
>= 6,694 < 9,167	>= 19,056 < 21,528
>= 9,167 < 11,639	>= 21,528 < 24,001
>= 11,639 < 14,111	

Figure 1 Population density in Rangoon

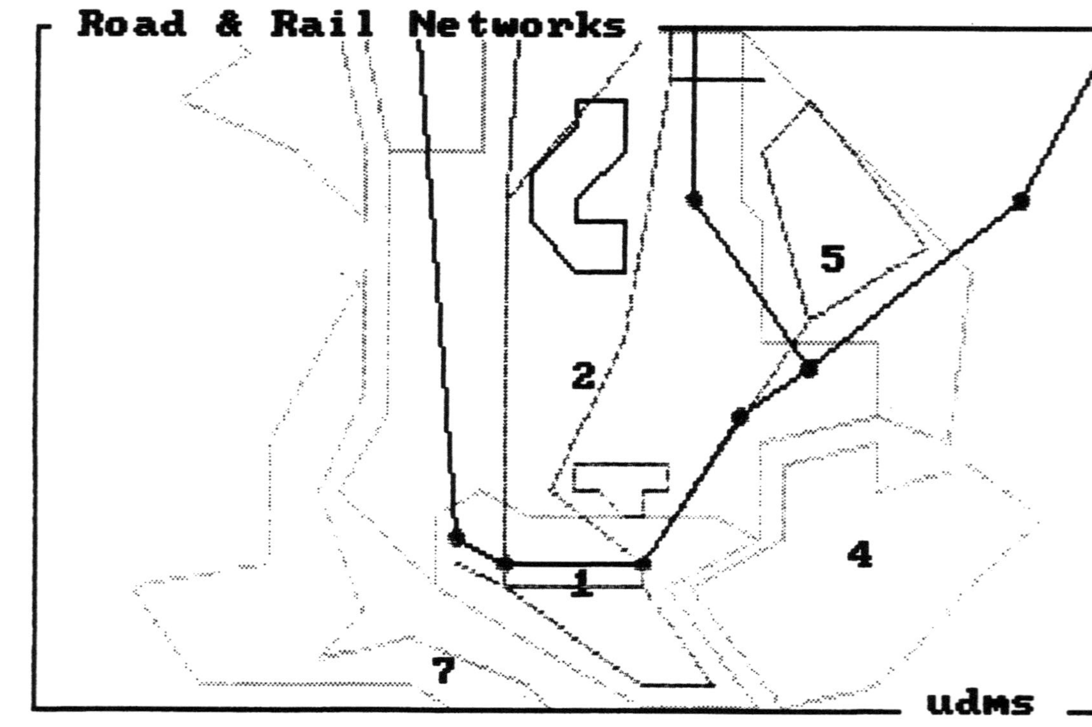

Figure 2 Highway and railway system in Rangoon

For the rest of the Housing Department, too, the range of possible computer applications is extensive. However, one particular type of application probably dominates all others in importance — and that is land and property administration.

The Land Department is the principal governmental agency responsible for land management in the Rangoon metropolitan area. Within this area and through its Land Survey and Land Revenue Sections, the Land Division is responsible for:

- Providing land surveys to public agencies and private individuals;
- Maintaining a register of land ownership and surveys;
- Assessing land rents and revenues;
- Collecting statistics on agricultural land use;
- Acquiring land and assessing land value; and
- Collecting ground rents and other land revenues.

The cadastral system has been in effect since 1894 and is apparently up to date — although standards and scales are said to be rather variable. It is estimated that there are about 20,000 registered plots in Rangoon and another 20,000 or so in the New Towns and elsewhere. Ownership is recorded on some 200 different cadastral maps, and transfers are recorded in a total of 37 Transfer Registers and 673 Town Registers. In principle, records are kept in local land offices for three years and then archived by the Division. In 1984, the Land Department reported handling about 60 transfers of ownership (or "mutations") per month.

In 1984, ground rent on all freehold land was BKs 7.00 per acre, with owners of less than one-quarter acre being exempted. Rents on leasehold land are subject to rental contracts between the lessee and the Land Division. Leases may be quarterly or annual or have terms of five, thirty, sixty or ninety years; and some leases are in perpetuity. Long-term leases normally give the Land Division the right to raise rents every five or ten years. Nevertheless, some leases are still based on pre-war rents of as little as BKs 2.00 per month (Housing Department, *The Housing Department: a Preliminary Evaluation* (Rangoon: mimeo, March 1984), pp. 21-4, supplemented by staff notes prepared for the author).

With all these records being kept manually, it is difficult to know to what extent public revenue from land in Rangoon could be increased, but there seems little doubt that a considerable increase is possible by collecting where there is now no collection, by following up on known arrears and by raising rents as and when the Land Division is entitled to do so.

At each step of the land-administration process, it is apparent that the Land Division could benefit from computerization: i.e., in maintaining the cadastre and preparing land surveys, in maintaining land ownership records and in maintaining leasehold records (including payment of rent and rent-adjustment dates allowed by the lease). The opportunities for both improving the standard of service and increasing public revenue are manifest.

Similarly, the Estate Management Division in the Housing Department is the principal governmental agency responsible for the administration of public housing. The Division administers an estimated 23,000 housing units, about three-quarters of them in Rangoon. The main functions of the Division are to allocate housing units to qualified tenants, to collect rents, to supervise and finance maintenance and generally to look after the welfare of tenants. The Division has 17 Estate Offices in Rangoon and another 12 in other parts of the country. Here, too, significant improvements could be made in collecting rents and in providing maintenance and other services by means of computerization. In this way, units could be efficiently allocated to families on the waiting list. Moreover, a complete and up-to-date record of the state of affairs in each unit could be maintained, including rent payments and expenditures on maintenance. Once again, it seems clear that there is plenty of scope for both improving service and increasing public revenue through computerization.

In addition to these urgent needs, computerization could help the Department in other ways. The departmental establishment comprises 1,095 staff: their records should be computerized in the interests of improved service to staff. Departmental accounts would also benefit from com- puterization, particularly insofar as control of cash balances, consolidation of branch-office accounts, and payroll (including payroll deduction) are concerned. The Planning Division could make use of computers for programme costing and programme evaluation, for keeping records of individual projects and works and for other data-processing functions. The Urban Water Supply and Sanitation Division could use computers for various kinds of data-processing (analysis of available resources and capacities), for engineering analysis and design (dams, spillways, distribution systems, etc.) and for forecasting demand and assessing pricing structures. No doubt, too, there are still other possible uses for computers elsewhere in the Department.

The Rangoon City Development Committee (RCDC) is the agency responsible for the management of urban services in Rangoon. It is divided into 15 departments and has over 5,000 regular employees and a further 1,500 "daily wage-earners". The functions and duties of the various departments of the RCDC are still largely based on the City Manual of 1922. Among the specific responsibilities of the RCDC are:

- assessment and collection of property tax;
- licensing of bazaars, markets, stalls, slaughterhouses, etc.;
- maintenance of roads, bridges, street lights, etc.;
- operation and maintenance of waterworks and sewerage systems;
- inspection of buildings and premises; and
- provision of parks and street-cleaning.

The RCDC is a self-financing institution; revenue collection forms a vital part of its operations. Nearly half its revenue derives from property tax, with the rest being made up of the various taxes, fees and fines it levies on the residents of Rangoon.

Once again, therefore, the opportunities for computerization are legion — word processing, accounting and financial analysis, database management, geographic information systems, engineering calculations, project management and so on. Just an operation such as maintenance of roads, involving street lights, vehicles, water pipes, sewers, electrical transformers and traffic signs, could be enormously improved through ready access to a system of up-to-date records about each individual item of plant.

The most attractive area for computerization is, once again, that of revenue collection because of the very tangible nature of the potential benefit (i.e., additional revenue). It is widely believed, for example, that improved collection of property taxes and licence fees could add tens of millions of Kyat annually to city revenues. (Needless to say, this would be quite enough to pay for an appropriate computer system.) But even in terms of "customer service", improvements could be expected from computerization. At the present time, for example, about 100,000 property-tax invoices are prepared each quarter by 60 "bill-writers" working manually. Even though two-thirds of the preparation time is devoted to checking for errors, many are inevitably missed, with consequent exasperation to customers as well as delays in payment.

In short, here, as in the Housing Department, the problem is not so much one of knowing whether computerization can help or even how, for there are many possible applications of computers to human settlements. Rather, the problem is one of knowing where and how to start with relatively limited resources.

Most of the applications described above can, in fact, be carried out on small and relatively inexpensive equipment. Of course, as the quantity of data gets larger or as the complexity of the processing becomes greater, using small computers becomes less and less convenient. But it rarely becomes impossible. Even in such cases, small computers are better than no computers at all. After all, large databases can always be divided into small ones. For example, information on leases or property-tax payments could be kept in a number of separate databases (e.g., one for each township), even though it might be most convenient to have it all in one large database. Similarly, computationally intensive processing can be planned so that it occurs overnight or on weekends. Again, it might be convenient if this did not have to be the case, but sacrifices are necessary when resources are limited.

The important thing is to make a start, even a modest one, so that users can become familiar with the potentialities and the challenges of computerization. Providing initial commitments are made compatible with the potential development of systems in the future, there is little to lose and much to gain by starting with relatively simple equipment and, as the skill and experience of departmental staff increase, building gradually from the ground up. For example, if it subsequently proves possible — through access to a powerful computer system — to establish a single tenant or property-tax database instead of many, nothing has been lost by having started out in a modest way. Indeed, depending on the circumstances, it may still make sense to carry out certain functions (e.g., data-entry) on small, simple machines and use the large system only for heavy-duty processing (e.g., printing notices or invoices).

There are two other good reasons for proceeding incrementally. First, user priorities will likely change as experience with computers grows. Users will see clearly what kinds of operations are suited to computerization and what kinds are not well suited. They may even think of completely new uses they had not previously imagined possible. Secondly, as time goes by, hardware prices tend to come down while performance tends to go up. In general, therefore, it is wise to postpone commitments to specific hardware until it is actually needed.

The principal conclusion of this case is that computerization for human settlements management in Burma is highly desirable for at least two reasons. First, computerization will help improve the standard of services that government provides to the people. Secondly, computerization will help increase the amount of revenue that government collects. As the World Bank has emphasised, the most critical factor in Burma's national development over the next few years is going to be its ability to mobilize its domestic resources for investment in productive activities (World Bank 1985, chap. II). Computerization can play a key role in helping to bring this about.

Case 6

Physical planning applications (Yemen AR)

Urbanization is a relatively recent phenomenon in Yemen, accounting for only just over 10 per cent of the roughly 8 million population. However, the urban growth rate (7.7 per cent) is now estimated by the World Bank at more than two and a half times the national growth rate (2.9 per cent). In the case of Sana'a (and perhaps other cities too), the population is widely believed to have already doubled since 1980. Moreover, 80 per cent of the urban population is concentrated in three cities: Sana'a, the capital; Hodeidah, the main port; and Taiz, the principal southern trading city and historic capital of Yemen. (Reference here and in the next few paragraphs is to the Staff Appraisal Report on the Second [Hodeidah] Urban Development Project (Report No. 4675-YAR), 9 November 1983, pp. 1–3.)

Cities in Yemen are ill-equipped to support this growth. In 1983, the World Bank characterized the situation in the following terms:

> The most striking feature of the urban setting in YAR is the lack of infrastructure . . . only about 40% of the urban population has access to potable water . . . . major cities still lack adequately designed and surfaced vehicular and pedestrian facilities . . . . most city centers experience heavy and nearly chaotic traffic conditions. . . . Garbage collection and disposal is limited, and education and health services are still at an early stage of development.

In particular, according to the Bank, the provision of adequate urban housing is constrained by three factors:

> Foremost among these is the absence of a formal land registration system [which has] substantially increased the uncertainty of tenure [and] undermined the establishment of an effective real estate taxation system.
>
> The second constraining factor lies in the high cost of [new] housing. . . .

> Third, access to housing, again particularly for low-income groups, has been further restrained by the absence of a well-developed housing finance system.

Finally, all of these factors are compounded (according to the World Bank) by the "limited" capacity of the existing institutional framework for dealing with them, as well as by the lack of adequate municipal revenues to finance and maintain new or even existing services. One way of strengthening institutional capacity and improving the collection and management of revenues is through improved data-processing — including the introduction of computerization.

The Ministry of Municipalities and Housing was established in 1979 to implement governmental policy and programmes with respect to urban development generally and housing in particular. Current plans to equip the Ministry with computers thus come partly as a logical outgrowth of the need for institutional development and partly as a reflection of the trend towards increased computerization within the government generally. Both technical and administrative activities within the Ministry can benefit from effective and efficient methods of data-processing. Such capacities are now readily available, thanks to developments in computer hardware and software over the past ten years. At the same time, computerization is becoming widespread in other parts of the government. Agencies are introducing computers into their operations and there are even prospects of electronic transfers of data between agencies. It is entirely appropriate, therefore, for the Ministry of Municipalities and Housing to be contemplating computerization at this time.

Almost any organization can benefit from computerization. So the question becomes not so much whether to computerize as how, where, when and how far to proceed. The key to knowing how to computerize (as anyone will tell you) is knowing why you are doing it: assess your needs and define your objectives; then design your computer system accordingly.

This is clearly good advice. However, even with the best of intentions, such advice is not always easy to follow. For one thing, if you are new to computers, you may not be able to correctly identify all your needs. There may be some cases where you do not fully appreciate the potential benefits of computerization. Similarly, there may be things you think you want to computerize but, when you realize how awkward and costly it is going to be, you may change your mind. For another thing, computers change the way you work. Things which were necessary in a manual system may no longer be necessary in a computer system. On the other hand, things which were too difficult or too time-consuming to contemplate with a manual system become quick and easy with a computer. To make matters more difficult still, computer technology is developing very rapidly and costs are continuously coming down, so that what is expensive and difficult today may be cheap and easy tomorrow.

For some organizations, things are made easy by the fact that they have one overwhelming computer need. For example, a utility company needs its billing computerized, an airline company its reservations system, a bank its accounts and so on. In other cases, however, computers can help meet a wide variety of needs; so deciding how best to go about it becomes complicated. This is the situation in which the Ministry of Municipalities and Housing finds itself.

Land-use and infrastructure planning

One of the responsibilities of the Ministry is land-use and infrastructure planning. This extends from the scale of an individual site to that of a whole city and includes the planning, design and execution of buildings, services and facilities. There are numerous cases where computers could help strengthen the capacities of the Ministry in this area.

For example, the Ministry requires and produces a great many maps at different scales. At the present time, most of these are prepared manually. However, once the necessary data are entered into a computer — and this can be done by hand, by means of a digitizer or by transferring the data directly from another computer — not only can a computer draw the corresponding map but it can also create a whole variety of other maps at different scales, covering different areas, showing different details, etc. Moreover, any numerical information on the maps (area, elevation, slope, rainfall, agricultural productivity, population, etc.) can be processed and displayed in different ways in order to reveal particular relationships or processes. A computer can even "overlay" one map on top of another and then draw a map showing the combined result. All of this can be done at a speed and with an accuracy that manual processing cannot match.

The Ministry also designs, builds and (in some cases) maintains a variety of urban infrastructure, including urban roads and bridges, public buildings and facilities, and street lighting networks. Here too computers can do a lot to facilitate the work involved by performing such functions as:

- Complex engineering calculations (such as stress analysis or seismic-risk assessment).
- Mathematical flow-models for estimating the capacity of sewerage and drainage systems, water-supply systems, etc.

- Computer-assisted design (CAD) of roadways, curbs, intersections, approaches, etc.; and of site plans, structural details, floor layouts, etc.
- Preparation of construction drawings, design charts, bills of quantities, etc. for project execution.
- Establishment of an inventory of completed projects (such as roads, bridges, street lights, market stalls and public lavatories) for purposes of service and maintenance, such inventories to include the location, type, age and service record of each facility.

All these and other aspects of infrastructure planning can be facilitated and enhanced through computerization. Indeed, it is probably no exaggeration to say that any one of these applications in the Ministry could pay for itself out of the cost-savings on a single project.

Services and facilities planning

In addition to planning for the design and implementation of urban infrastructure, the Ministry is also involved in the supply of housing, transportation and other services and facilities. The Ministry is particularly concerned about meeting the needs of low-income groups. Here too, computers can play an important role in helping Ministry officials carry out their responsibilities.

For one thing, planning for services such as housing and transportation requires the collection and analysis of information about how individuals behave and what kind of environment they want to live in. This means household surveys, roadside interviews, consumer questionnaires and so on. It is possible to record and analyse by hand the large quantities of data that survey research typically produces, but it is a great deal easier to do so with the help of a computer and suitable statistical software. Inevitably, the easier something is to do, the more likely it is to be done. Thus, computers can help make sure that when data are gathered for planning, they are properly and fully used.

For another thing, computers make mathematical modelling easier to do than it is by hand. This is not because models are easier to create or more accurate when done on a computer, because often they are not. It is not even because models involve a lot of tedious mathematical calculations, although often they do. Rather, it is because the value of a model derives from the fact that it can be "run" again and again to test the effect of different assumptions and different parameters. Here is where the computer can really help: once a model is entered into a computer, the model can be run repeatedly with little effort (and no loss of accuracy). This means planners can identify various options and carefully examine the consequences of each. To give a few examples:

- The World Bank has developed a computer model (the Bertaud model) specifically for sites-and-services schemes. With this model, planners can test the effect of various assumptions on a particular site. For example, the computer can show the effect on the physical design of the scheme of different plot sizes, different road networks, different public-space allowances etc. Then, when it comes to assessing affordability, the model can assist in analysing the effect of different assumptions about loan terms, downpayments, repayment schedules, minimum household income, etc.
- Similarly, there are computer programmes for estimating road transportation demand, using conventional techniques of trip generation, trip distribution (using a gravity model) and traffic assignment. Such models allow planners to examine the effectiveness of various road patterns in meeting transportation demand as well as assess the extent and impact of additional traffic loads created by urban developments (public or private).
- Thirdly, long-range planning inevitably depends on population forecasts. There are various techniques for making such projections, but there is (of course) no foolproof way of knowing what the future will be. What the computer can do is quickly and accurately present the results of using any or all of the forecasting techniques under various assumptions, so that planners can concentrate on assessing which approach and which assumptions are most likely to reflect the particular situation they are studying. Computers can be used in this way for both small-area projections (using shift-share analysis, for example) and for regional and national projections (using cohort-survival techniques, for example).

Essentially what computers do in all these cases is to take some of the drudgery out of planning. This is more than just a matter of a planner's convenience. It has to be recognized that someone who has just spent the best part of a week designing a housing site or calculating a population projection is far less likely to be disposed to question its methods and assumptions than if he or she had created the model by simply pressing a few keys on a computer. In other words, in removing some of the tedium from the planning process, computers do not just make it faster; they should help improve the quality of planning as well.

Land management

Although land registration on a national basis is constitutionally the responsibility of other agencies of government, the Ministry has become involved *de facto* in land administration through its various sites-and-services schemes. Originally, Yemen relied entirely on private conveyancing for establishing land title. There was an attempt (Law No. 12 of 1976) to create a system for voluntary registration of title; but this was superseded by Law No. 13 of 1977 which provided for compulsory registration of any new transactions involving urban land and of title deeds pertaining to property for which either building permits or mortgages were required. (Law No. 13 also restricted individual land holdings to 50 *libna*, or about 2,200 square metres, and imposed a flat 2 per cent land transfer tax.) However, none of these changes has resulted in the kind of land registration system that has been called for in a succession of studies made for the government: namely, a fully fledged system for registering and endorsing title along the lines of a Torrens system (Hughes Report 1983).

Yet, in its sites-and-services schemes, the Ministry has found it desirable to provide (albeit on a very small scale) just such a system. A complete land register of its projects provides the Ministry with an up-to-date and unambiguous record of ownership for every plot in a scheme, thus eliminating doubt and controversy. For the new owner, the land register provides security of tenure. For the creditor (in this case the Ministry), the land register provides security for his loan. These advantages are not transitory; indeed, as time goes by, and the original owners and creditors are replaced by new ones, the security provided by the land register will become increasingly important. The government is also well aware that land registration systems developed by the Ministry for its own projects can serve as prototypes for an eventual land registration system for the whole country — another reason for the Ministry to develop an efficient and effective system.

There are at least two special ways in which computers can contribute to a land registration system, one administrative and the other technical. From an administrative standpoint, a computer can provide an electronic catalogue of and index to information contained in the land registry. The advantage of using a computer is that updating, sorting, searching and analysing the database become much easier and quicker than they are by hand. For example, in most manual systems, the only way to get title information for a particular plot is by its plot number, whereas with a computer you can search on the basis of any known information. Similarly, if you want to know something about all plots that have changed ownership in the past six months or all plots over a certain size, the computer can quickly read through its records and provide it.

The second way in which a computer can support a land registration system is at a technical level. Calculating plot boundaries and locations from field data involves a lot of repetitive calculations, which a computer can easily be programmed to do. From these data, moreover, a computer can then quickly provide a survey drawing of the registered property. Indeed, if the numerical description of the property is itself stored in the computerized registry, the computer can produce copies of the survey whenever they are required. As the 1983 Hughes Report explains,

> Land registration . . . produces a lot of data. The ideal way to handle data today, both geographical and written, is by electronic data processing (EDP). . . . Its introduction may start in a modest way . . . by sorting and tabulating data and providing lists of records. Simple equipment would be used. Progress could [then] be made through to the technical stage of doing cadastral computations, computer-assisted drafting and the administrative work. EDP at this stage offers the advantage of speed, accuracy, accessibility and security.
>
> (Hughes 1983, page 33)

Plot and loan administration

As a byproduct of its sites-and-services schemes, the Ministry has also become involved in selection of beneficiaries and administration of a loan scheme to provide them with construction capital. In Sana'a, there are some 2,000 such beneficiaries, in Hodeidah another 3,000, and the Third Urban Development Project may involve as many again. (Although the Housing Credit Bank has been operating a similar scheme for about 1,000 medium-income to high-income households since 1978, the government has again preferred to let the Ministry implement its own programme as a possible prototype for a national programme.) There are three key steps in this process: selecting beneficiaries and allocating plots; preparing documentation for each plot and loan; and monitoring loan repayments. Computerization can facilitate each one of these activities.

With a computer, the selection of beneficiaries can be done in a variety of ways. For example, selection can be based on the successive use of a series of criteria, until a sufficient number of applicants have been chosen. Alternatively, a system can be devised for assigning "points" to applicants on the basis of certain factors; then, the computer works out the "point" scores for each applicant and produces a list of applicants ranked according to their scores. Still another method is to instruct the computer to select beneficiaries on a completely random basis.

Much the same options are available for the allocation of plots to beneficiaries. Allocation can be essentially deliberate, according to some set of priorities or system of "weights"; or allocation can be done "blind" by using random numbers. The Housing Bank in Jordan, for example, uses its computer to support a policy of allocating units according to certain criteria. The Delhi Development Authority in India, on the other hand, allocates its housing units by computer in order to ensure that every beneficiary has an equal opportunity of being allocated any particular unit.

Once beneficiaries are selected and assigned plots, there comes the process of preparing documentation for the title and (if required) the loan. Each case involves numerous documents, legal and financial, almost all of which are more or less standardized in form and content. Computerization can facilitate the preparation, monitoring and storing of such documents.

Finally, there is the process of loan administration — maintaining accounts for each individual loan, recording payments against an agreed schedule of repayments, adjusting loan schedules to reflect changes in the rate or term of a loan, identifying accounts in arrears, calculating the overall position of the loan fund, projecting likely cash flow and preparing other reports and summaries. Here too, computerization can make the job more efficient, more accurate and more effective than if it is done manually.

Project management

Almost all the Departments in the Ministry are involved in letting and supervising contracts; so project management is one of the important activities of the Ministry. There are several discrete steps involved here, and once again computers can do a lot to assist officials with their work.

One of the main steps in the process is preparation of first tender and, then, contract documents. Many of these documents are quite similar to one another or, at least, incorporate many of the same standard paragraphs and clauses. Furthermore, each document invariably goes through several drafts, often involving only small textual changes, as bids are received and contracts finalized. The result is that substantially the same text and substantially the same drawings have to be prepared over and over again, with only small changes. Each time this occurs, the revised version needs to be checked not just in its revisions but in its entirety, in case new errors have crept in.

With a computer, however, text and drawings need be prepared only once and then stored electronically. Then, they can be revised and edited as required and reproduced with the confidence that, once they are correctly entered in the computer, it will, thereafter, reproduce them correctly every time.

Another critical step in project management is monitoring results and evaluating performance against a schedule, so that it becomes clear when a project is falling behind and what needs to be done about it. There are several useful techniques for doing this, including the simple but effective "critical path method" and the sophisticated "Program Evaluation and Review Technique" originally developed for monitoring the US Navy Polaris submarine programme. However, often these methods are tedious to apply manually, particularly when the project is complex and the deviations from schedule numerous; so simple indicators are used, such as draws against budget or outputs produced. The result is that Ministry officials may not be aware of how serious some delays may be in terms of the eventual completion date nor be in a position to advise contractors how to reallocate their resources so as to bring a project back on schedule. With a computer, on the other hand, sophisticated project management takes little more time than is needed for entering the necessary information in the computer. Critical-path calculations can be made time and time again to show the effect of different resource allocations; the corresponding project costs can be estimated and re-estimated; even the wall-size Gantt charts can be drawn and re-drawn to reflect changes in project planning. With a computer, effective project supervision takes only minutes.

Similarly, computers can be very helpful when it comes to keeping project accounts. Officials will find it easy with a computer to monitor draws against budgets and to allocate expenditures to specific budget lines. Computers can also help in analysing bills of quantities and verifying contractors' invoices. In this way, secure in its understanding of the status of a project, the Ministry can enter into a frank and productive dialogue with its contractors regarding costs, performance and schedules.

In these various ways, computerization can provide Ministry officials with a continuous, accurate and up-to-the-minute picture of the status of every project under way. This would include information on physical performance, such as results to date and projected completion date, as well as financial information, such as expenditure to date and projected cost overruns (if any).

Administration and finance

In concentrating (as this case has done) on how computers can aid the Ministry in discharging its line responsibilities, it is important not to neglect the internal operations of the Ministry itself. As with any organization, the Ministry depends for its success on the efficient and effective management of its resources, including personnel and budget. Here too, computers can play an effective part in enhancing the operation of the Ministry.

Among the principal uses to which computers could probably be put in the administration and management of the Ministry are the following:

Computers are highly effective as word processors, especially where standard text (as in legal documents), numerous revisions (as in reports for official publication) or form letters (as in correspondence with the public) are involved.

Computers are ideal for keeping accounts, preparing and monitoring budgets (whether for projects or for municipalities) and compiling periodic reports for management.

Computers can be used for maintaining personnel records and preparing a payroll (including calculation of bonuses where applicable); or for keeping records of suppliers and contractors for procurement and/or paying bills.

Computers can be used for keeping an inventory of assets (such as office equipment, library books, vehicles and real property) or for managing stores to maintain an efficient level of stock (e.g., of office supplies or project equipment).

Computers can be used for improving management of resources, such as vehicles (including service schedules, operating costs, utilization rates and driver assignments) or office space (including fixed costs, maintenance costs, departmental assignments and occupancy rates).

Computers can be used for preparing publications and press releases, including editing, layout, illustration and even typesetting.

No doubt there are other potential uses for computers in respect of administration and finance in the Ministry, but those mentioned here will give some indication of how and where computerization could help.

Research and policy analysis

The use of computers for research and policy analysis is, in a sense, both a reflection and an extension of some of the points discussed above. Essentially, policy-making requires the examination and evaluation of possible courses of action in the light of different assumptions about the nature of current and future conditions. Here too, computers can assist senior management in the Ministry.

For one thing, computers can help managers monitor current conditions and performance effectively. The more computers are used within the Ministry, the easier it will be to improve the flow of information from the field to headquarters and from lower to higher levels of management. With effective computer systems in place along the lines indicated above, it should not be difficult to provide quick and up-to-date answers to questions such as the following: what is the status of this particular project? what tolerances were used for that particular bridge? how many housing units were built last year? how many housing loan accounts are in arrears this month? how many street lights are there in Hodeidah and what is the annual cost of maintenance? how much is the Ministry spending on vehicles and how intensively are they being used? Even without a full-scale computerized management information system — which would have to be regarded as a very ambitious undertaking at the present time, at least until the Ministry has had experience with computers — improved management information should nonetheless be a byproduct of computer use at the operating level.

The second main contribution that computers can make to research and policy-making has to do with data analysis and modelling. There are many different ways in which this can occur; suffice it to give here a couple of examples:

> The Agency for International Development has developed a computer model for assessing housing needs based on 27 commonly available input variables. In some cases, all that is needed is a single value for the whole country in the base year; in other cases, the model allows for distribution of the variable over time, by region and/or by income group (quintile). Using these data, the model performs a series of calculations designed to determine what housing needs are expected to arise over the next 20 years and how far they are likely to require public subsidy. Naturally, the model can be run as many times as required to examine the effect of using different initial assumptions or implementing different programmes of public subsidy.
>
> Cost-recovery is another important policy area for the Ministry. Computers can be very useful in estimating the capital costs of new infrastructure, evaluating the flow of benefits, determining appropriate user charges and assessing the fiscal impact of such charges. With this in mind, the Bureau of Economic and Business Research at the University of Florida has produced a collection of computer programmes (spreadsheet templates) for local government which include the following functions:
>
> - Estimating the capital costs of water treatment
> - Estimating the capital costs of water distribution
> - Estimating the capital costs of wastewater collection
> - Estimating the capital costs of wastewater treatment
> - Estimating the capital costs of non-specific services

- Financing options for capital improvements
- Determining solid-waste disposal charges
- Fiscal impact analysis — residential
- Fiscal impact analysis — non-residential.

This collection is intended for use in the USA; so some of the programmes would need modification before they can be used in other countries. Nevertheless, they can still provide a valuable indication of how computers can facilitate the evaluation of different policy options for achieving cost-recovery.

In the end, of course, computers do nothing that humans cannot do, but the fact that computers do some things more quickly and efficiently than humans means that management can expect significant improvements in productivity from the use of computers. Not only will planners be able to shift some of their tedious work to the computers; but the latter will be able to do more of it in less time. This will have the added advantage of freeing planners from routine activities, thereby allowing planners time for the creative aspects of their work which are so essential when it comes to policy-making.

From detailed project management to national policy analysis, it is clear that opportunities for computerization abound in the Ministry of Municipalities and Housing. Potential applications include text processing, mapping, database management, accounting, financial analysis, statistics, mathematical computation, graphics, physical design, modelling and simulation. There is no shortage of opportunities here. Indeed, it is precisely this abundance of opportunities that leads to a recommendation that the Ministry adopt a computer strategy characterized by a limited financial commitment at the outset but providing widespread decentralized access to computer facilities. Such a strategy will give the Ministry the experience it requires to be able to identify its long-term computer needs and priorities.

The best way of meeting these objectives is probably by means of stand-alone microcomputers. The advantages of this approach are that:

Microcomputers are powerful enough to be able to provide most of the capabilities currently required by the Ministry. There is a lot of off-the-shelf microcomputer software available to help meet the Ministry's needs, some of it already in Arabic as well as English.

Microcomputers are relatively easy for non-specialists to learn how to use, thanks to the variety and sophistication of the available software.

Microcomputers are flexible and portable, so they can easily be moved from one office to another as required.

Microcomputers provide redundancy in the event of equipment failure. If one microcomputer fails, it does not affect any of the others.

Microcomputers are cheap enough for access to be provided on a widespread basis at a reasonable cost.

Microcomputers do not need expensive support facilities, such as air-conditioning, air filtration and dehumidifying units. In addition, parts and service are not as expensive as they are for large computers.

Microcomputers can share data and software just by physically exchanging disks. Electronic communication by modem or local area network (LAN) is also possible at relatively low cost.

For these and other reasons, microcomputers represent a sensible first step towards computerization.

Furthermore, even if a powerful, centralized system is subsequently acquired by the Ministry, that will not make the microcomputers obsolete or redundant. Microcomputers will still be able to meet certain needs more efficiently than the new system (depending on exactly what type of new system is acquired). Alternatively, microcomputers can always serve as terminals for a new system. In any case, microcomputers will have promoted computer "literacy" within the Ministry to a degree that a central computer could never have done for the same cost.

There is a tendency (which vendors do little to discourage) to think that computer purchases should be designed so far as possible to meet all future needs, but this can result in considerable "over-specification". Besides, with technology changing as quickly as it is and with prices coming down all the time, the prudent course is often to postpone purchases until they are really necessary. In short, it is important to recognize that this first computer purchase by the Ministry is not going to be its last. Thus, it is best to buy computer equipment on the assumption that demand will eventually outgrow supply than it is to try to ensure a persistent over-supply. In the first place, the Ministry will have the satisfaction of knowing that its initial investment is being put to good use; and users' complaints about the shortcomings of the initial equipment will provide a good guide to exactly what additional capacity is required. In the second place, in spite of spending a lot of money, the Ministry may still find that it has failed to anticipate every need. Worse still, the Ministry may find itself in the embarrassing position of not being able to use all the capacity it has.

Equipment and software, however, represent only part of the process of computerization. There are at least three other vitally important factors. These are: commitment, training and applications. Learning to use a computer — even a small and "friendly" computer like a micro — requires an effort. If people are going to make this effort, they need the motivation and the opportunity to do so.

First of all, senior management has to make clear that it is committed to computerization in the Ministry and that it is prepared to support staff members who start to move in this direction. One of the best ways to do this is to appoint a competent person to be full-time "Co-ordinator of Microcomputers" in the Ministry and to encourage one or two staff members in each Department to play a similar role there, even if in an informal capacity. The purpose of this appointment is not to create a "Computer Department" *per se* within the Ministry but rather to provide users with a quick and convenient source of help while they are learning how to use their computers. Even if the Co-ordinator does not have the skill or the resources to solve all the problems that arise, he or she will at least be able to advise what to do. What is most important, the Co-ordinator will represent a tangible expression on the part of senior management of its commitment to the process of computerization.

Another vital supporting factor in computerization — and one that can hardly be over-emphasised — is training. Although most computer programmes have comprehensive (if not always very readable) manuals and (in some cases) computer-based "teach yourself" tutorials, people still need a certain amount of formal training. This is particularly true at the initial stages. Given a good introduction to computers and to a particular type of software, most users find they can improve their skills on their own (again, as long as they have the motivation and the opportunity to do so) but they do need help in starting out and encouragement in continuing. The Ministry will find it worthwhile to provide for some staff training as soon as possible after delivery of the computers.

Finally, computerization depends not just on having a computer system and training people to use it. Staff members must also be able to perform efficiently tasks which are useful to themselves and the Ministry. In other words, specific computer applications have to be developed. In some cases, such applications have to be custom programmed using a programming language such as BASIC or FORTRAN; but most of the Ministry's current needs can probably be met by adapting "generic" software (such as database management systems, spreadsheets, accounting packages and computer-assisted design programmes). In time, Ministry staff members will undoubtedly develop their own skills in applying such programmes to their needs. In the initial stages, however, the Ministry will find it worthwhile to provide for some staff assistance in developing software applications specific to its needs.

Case 7

Social planning applications (Malaysia)

The Ministry of Housing and Local Government is the principal agency of the Malaysian government responsible for national policy on human settlements. Inasmuch as powers over land and local government are assigned by the Constitution to the state governments, however, the role of this Federal Ministry is primarily one of co-ordinating and providing technical support for various state activities. Where the Ministry is involved in implementation, it acts at the request of and on behalf of the respective state governments.

Under the Fourth and Fifth Malaysia Plans (1981–5 and 1986–90 respectively), the Ministry has three objectives. One is to ensure that every Malaysian, especially those in the low-income groups, has the opportunity to own a house of satisfactory quality. Thus, in addition to maintaining a continuous watch over the national housing situation, the Ministry puts a lot of emphasis on development of low-cost housing as well as renovation of existing housing stock in rural areas. During the current Plan period, the Ministry is expected to supervise construction of a total of 45,800 low-cost public housing units financed by federally subsidized loans to the states (see Table 5). Moreover, under a special "crash" programme over the next three years (1986–8), the Ministry is also to be responsible for overseeing construction of an additional 80,000 low-cost housing units per year; these units will be financed by commercial loans backed by the federal government. Through the National Housing Department, the Ministry has nearly 300 officers in various regional offices to assist with implementation of housing projects.

The second role of the Ministry is to encourage the efficient organization and functioning of local-government authorities throughout the country. Since the restructuring of local government in Malaysia in 1973, there are 93 local authorities, 15 municipal councils and 78 district councils. Local councils are creatures of and responsible to their respective state governments. However, the federal government through the Ministry of Housing and Local Government is responsible

Table 5 National housing programme for the period of the Fifth Malaysia Plan, 1986–90

	Type of Housing Unit			*Total*
	Low-Cost	*Med-Cost*	*High-Cost*	*Target*
	(Housing units)			
Public Sector				
Public housing	45,800			45,800
Land development schemes	57,500			57,500
Staff housing	4,400	22,500	100	27,000
Other programmes	13,200	5,400	100	18,700
Total public sector	120,900	27,900	200	149,000
Private Sector				
Private developers	370,400	146,000	23,600	540,000
Cooperative societies	3,700	6,300	2,500	12,500
Total private sector	374,100	152,300	26,100	552,500
Total Both Sectors	495,000	180,200	26,300	701,500
Average housing units per year	99,000	36,040	5,260	140,300
Type of unit as a percentage of total	70.6	25.7	3.7	100.0

Source: "Malaysia's Housing Programme, 1986-90", Table provided by the National Housing Department (Kuala Lumpur: photocopy, no date).

for advising and assisting local authorities in respect of planning and management and for providing funds and technical assistance for capital projects and other development activities.

The third role of the Ministry is to promote town and country planning throughout the country. Here too, while the control and use of land is constitutionally assigned to the states, the federal government plays an important role in providing the human and financial resources required for effective planning. The Ministry is responsible both for preparing a national physical plan and for assisting local authorities in drawing up appropriate municipal and district plans. Moreover, the Ministry, through the Town and Country Planning Department, provides some 500 officers on secondment to planning departments in the state governments.

Some departments in the Ministry have been making limited use of computers for several years, but it is only recently that the Ministry has acquired some equipment of its own. The Ministry currently has a total of six microcomputers and one dedicated word processor. The microcomputers are of various kinds and capacities and are located throughout the Ministry. In summary, the situation is as follows:

- There are two independent (and incompatible) microcomputer systems in the National Housing Department, where they are used more or less exclusively by that department.
- There are two basic PC systems: one is an IBM machine located in the Planning and Development Division, where it is used jointly by that division and the New Villages Division; the other is an IBM-compatible machine located in the Local Government Division and used exclusively by that division.
- There are two IBM PC/AT systems: one is located in the Research Division, where its use is shared with the Town and Country Planning Department; and the other is in the Licensing Division, where its use is shared with the Enforcement Division.

In general, equipment is being used primarily for word processing (with Wordstar) and database management (with dBaseIII). Two other software packages are also available in the Ministry, namely, Lotus 123 and AutoCAD. The first is already being used in some parts of the Ministry; use will undoubtedly expand as experience and access to equipment increases. AutoCAD is not yet in widespread use, partly because of its relatively specialized applications and partly because the Ministry does not have any appropriate input/output peripherals (such as a digitizer to help enter graphic data or a plotter to print out maps or drawings).

Among the computer applications currently in operation or close to implementation within the various units of the Ministry are the following:

1. Contractors' licences database

This database is being prepared by the Licensing Division using dBaseIII and consists (so far) of two files. There is one file (256 bytes) on contractors' names and addresses; it contains records on about 2,500 contractors. The second file (475 characters) contains data on specific projects; it has about 650 records in it so far and is to be backdated to 1981. Among other uses, these files will be helpful for monitoring (and printing address labels for reminders about) the progress reports (Form 7F) which contractors are required to submit every six months.

2. New villages database

This database is being prepared under dBaseIII by the New Villages Division and will consist of five separate data files on the 492 "new villages" still extant; total size of the database when completed is expected to be about 0.5 Mb.

3. Gombak pilot project

This is a database on one particular district of Malaysia compiled by the Local Government Division as part of a UNESCO-sponsored development project.

4. Building materials catalogue

This database is a project of the Research Division and is still only at the planning stage. The idea is to prepare and keep up to date a comprehensive list of the characteristics, standards and costs of all the building materials currently in use in Malaysia. This database would be made accessible not just to Ministry officials but also to architects, builders, etc. in the private sector.

5. Housing prices index

Based on computer analysis with Lotus 123, the Research Division publishes a periodical index of housing prices in Malaysia for different types of housing in various parts of the country.

6. Quantity surveying system (QSS)

This is a system for managing and monitoring building contracts based on a commercially prepared, multi-user package. Among its capabilities are: preparation of bills of quantity (BOQs), maintenance of project accounts and project monitoring. The system is used by the National Housing Department for contracts which it supervises on behalf of the various state governments.

7. Low-cost housing project masterlist

This is a dBaseIII database on the approximately 550 low-cost housing projects (representing about 170,000 units) that are currently under way in Malaysia. The database is maintained by the Planning and Development Division in consultation with the National Housing Department and the Licensing Division.

In general, it would appear that the Ministry has made a sound, if somewhat modest, start on the process of introducing electronic data-processing into the work of its departments and divisions. As far as can be judged, the existing equipment is being put to good use, and those uses are becoming increasingly sophisticated as staff members gain in skill and experience. Attitudes towards computerization seem positive throughout the Ministry. In short, it can hardly be doubted that the Ministry's investment in computers to date will be fully recovered through increased productivity within a year or two. However, the scope for further productivity gains is still enormous.

The key to the success of the Ministry's approach so far has been its decentralization. Even though current computer resources are quite limited, the Ministry has nevertheless spread them out among its various departments and divisions in order to allow as many uses and as many users as possible to take advantage of computerization. Furthermore, the Ministry does not appear to have tried to exert any central influence or control over how and where the equipment should be used. The result is that each unit has been able to apply its computer resources to its own particular requirements in the ways it judged most appropriate. Of course, microcomputers are ideally suited to this kind of strategy, for they are by nature independent, flexible and easy to use. Furthermore, because of the rapid pace of technological change and the "open architecture" of most microcomputer systems, it is no longer necessary or even wise to "buy ahead" of current needs the way it once was.

Thus, the Ministry of Housing and Local Government has its own peculiar set of needs, and their chief characteristic is not so much standardization, high speed or simultaneous access as it is diversity of applications. Of course, the Ministry can and will achieve some degree of standardization, some improvement in speed of processing and some sharing of data through computerization; but none of these is as important as maximizing the number and variety of computer applications in the Ministry. Table 6 gives an indication of just how extensively microcomputers could be used in the various parts of the Ministry.

One of the key factors in running diversified computer applications is training. Training plays an indispensable part in learning about computers, as indeed in learning about anything. Microcomputers have always been intended for use by non-specialists in a way that large machines never were. Microcomputers are meant for ordinary people — secretaries, engineers, planners, administrators, etc. Thus, it is not necessary to be a computer scientist or a computer programmer to use a microcomputer — just as you can learn to drive a car without being an auto mechanic or a racing driver. Thus, most current users in the Ministry are in fact self-taught, with help from others when they started and occasionally now when they find they need it.

Formal training courses are useful for overcoming initial fears about computers and for getting people started on particular software packages, but after that what most users need is individualized help. They need a rapid source of advice for dealing with problems encountered on the job. The reason is that, for most people, using computers is not an end in itself; it is just a means for improving their professional effectiveness. Thus, most people do not want and do not need to perfect their computer skills for their own sake; all they want is to be shown how the computer can help them do their jobs and then to be able to get help when they need it.

Table 6 Selected microcomputer uses and their potential applications in the Ministry of Housing and Local Government in Malaysia

1. WORD PROCESSING (WP)

POTENTIAL USERS:	All units in the Ministry.
SPECIAL HARDWARE:	None required, although the Ministry may eventually want to make use of high-quality laser printers for producing documents of typeset quality.
SOFTWARE SOURCES:	Numerous commercial and non-commercial WP packages are available, including some programmes with features such as the ability to create customized form letters (usually called "mail-merging"), to perform simple arithmetic operations, to provide limited text-management features (such as sorting) and to access spelling-checkers and/or thesauruses. There are also (usually separate) packages for page layout and graphics on laser printers.
COMMENTS:	Most units in the Ministry already have some experience with basic WP. However, elaborate applications (such as mail-merge and use of standard "boiler-plate") are not yet in extensive use.

2. DATABASE MANAGEMENT SYSTEMS (DBMS)

POTENTIAL USERS:	All units in the Ministry.
SPECIAL HARDWARE:	None required, although large databases (from 200 Kb up) are fastest to retrieve and easiest to manipulate if a hard disk is available. (Hard disks should in turn be provided with a suitable backup system). The entry of prepared data can also be significantly speeded up by use of an optical character reader (OCR).
SOFTWARE SOURCES:	Numerous commercial and non-commercial DBMS packages are available. dBaseIII and Rbase 5000 are widely used commercial packages with their own programming languages and limited "relational" capabilities. To optimize the use of such complex packages, staff may find it useful to acquire "utility" packages for helping prepare screen/report layouts and compilers to speed programme execution. For simple tasks, on the other hand, staff may prefer a simple file management programme (pfsFile) or an "analytical" DBMS (like Reflex).
COMMENTS:	dBaseIII is already in use in the Ministry; however, there are numerous potential DBMS applications still unrealized, owing to lack of access to equipment. Among these are: planning and project databases in various units of the Ministry; and matters such as personnel, payroll, stores, supplies and inventory in the Administration and Finance Division and other units.

3. ACCOUNTING, BUDGETING AND FINANCIAL ANALYSIS

POTENTIAL USERS: Administration and Finance Division; Local Government Division; New Villages Division; Research Division; Housing Loans Scheme Division; and others.

SPECIAL HARDWARE: None required, although again a hard disk may prove useful for larger files. (However, note that most spreadsheet programmes operate entirely in memory; so maximum file size is limited by available RAM.)

SOFTWARE SOURCES: Straightforward budgeting and financial analysis can be done with a standard spreadsheet package (e.g., Lotus 123, Supercalc or their non-commercial counterparts) or even a database management programme. However, complicated accounting operations (such as payroll, purchasing or accounts payable) may eventually require specialized software, probably from commercial sources.

COMMENTS: Among the many potential applications here are several in the Administration and Finance Division (e.g., maintenance of the Vote Books, recording the schedule on Vouchers, tracking Purchase Orders, maintenance of expenditure records, and recording allocation warrants issued under the Development Budget); as well as project budget monitoring in the National Housing Department, the Local Government Division, the New Villages Division and elsewhere.

4. STATISTICAL ANALYSIS

POTENTIAL USERS: Town and Country Planning Department; New Villages Division; Local Government Division; Enforcement Division; Research Division; and others.

SPECIAL HARDWARE: None required but: (a) some of the sophisticated commercial software requires a hard disk; and (b) for large data-files it may be desirable to add a "mathematics co-processor" (such as the Intel 8087 or 80287) in order to reduce processing time.

SOFTWARE SOURCES: Commercial and non-commercial packages are widely available, although (in this case) the latter do not usually offer the same level of performance as the former. Two of the most widely used commercial packages are SPSS and SAS. There are also ready-made templates for spreadsheet packages to handle simple or specialized requirements. The United Nations Fund for Population Activity (UNFPA) has (non-commercial) microcomputer software especially for demographic analysis.

COMMENTS: Extensive questionnaire survey projects are planned or already under way in the Town and Country Planning Department, the New Villages Division and the Local Government Division. These are obvious candidates for computerized statistical analysis.

5. COMPUTER-ASSISTED DESIGN (CAD)

POTENTIAL USERS: Town and Country Planning Department; National Housing Department; Research Division; and others.

SPECIAL HARDWARE: Effectively requires special peripheral equipment for data input (e.g., a digitizer) and output (e.g., a plotter). Enhanced colour capabilities are also an obvious advantage for this kind of application. A mathematics co-processor will also speed up computationally intensive operations (such as rotation or zooming).

SOFTWARE SOURCES: Full-scale computer-assisted design (CAD) software is largely commercial, although prices tend to vary more than differences in capabilities would seem to warrant. AutoCAD is probably the most widely used commercial package, but there are packages with extensive features (such as full three-dimensional capacity). On the other hand, there are some simple but still very useful programmes available, including drawing programmes such as Dr Halo (commercial) and PC-DRAW (non-commercial); or the famous sites-and-services planning programme (the Bertaud Model) developed for and distributed by the World Bank and other international agencies.

COMMENTS: For serious engineering and architectural design work, there are also dedicated CAD minicomputer systems available. However, for its purposes, the Ministry will probably find that microcomputers can provide an impressive level of capability, especially when allied with appropriate peripheral equipment.

6. MAPPING, SURVEYING AND GEOGRAPHICAL INFORMATION SYSTEMS (GIS)

POTENTIAL USERS: Town and Country Planning Department; National Housing Department; Local Government Division; Research Division; and others.

SPECIAL HARDWARE: Here too, special peripheral equipment for data input (e.g., a digitizer) and output (e.g., a plotter) is an advantage. Enhanced colour capabilities are also useful.

SOFTWARE SOURCES: There are some commercial programmes for mapping and GIS available for microcomputers (e.g., PC-MAP); moreover, the United Nations Centre for Human Settlements (Habitat) has a non-commercial package called Urban Data Management Software (UDMS). Standard survey calculations and the drawing of simple property surveys for housing plots can be programmed easily enough in BASIC or FORTRAN, if those languages are made available.

COMMENTS: Full-scale GIS capabilities really require a dedicated minicomputer system. The Ministry will have to decide what level of capability it requires and be guided accordingly.

7. MUNICIPAL ADMINISTRATION AND MANAGEMENT

POTENTIAL USERS: Local Government Division; Administration and Finance Division; Research Division; and others.

SPECIAL HARDWARE: None required.

SOFTWARE SOURCES: In view of the advantages of promoting uniformity among the practices of different local authorities, popular generic software packages (dBaseIII and Lotus 123 or similar) are probably most suitable.

COMMENTS: There is a whole range of potential applications here — such as land registration and taxation, municipal budgeting and accounting, licensing and revenue collection — that could benefit from computerization. Not only would this help local authorities to function more efficiently, but, since data on diskettes can readily be transferred from local authorities to the Ministry, it would do a lot to help the Ministry get accurate and up-to-date information about local conditions.

8. LOAN ACCOUNTS MANAGEMENT

POTENTIAL USERS: Housing Loans Scheme Division; Administration and Finance Division; and others.

SPECIAL HARDWARE: None required, although large databases (from 500 Kb up) are fastest to retrieve and easiest to manipulate if a hard disk is available. (Hard disks should in turn be provided with a suitable backup system.)

SOFTWARE SOURCES: The requirements of the Ministry can probably be met with some fairly simple packages or even a customized application of a DBMS programme (such as dBaseIII). There is also a Housing Finance Software (HFS) package available on a non-commercial basis from the United Nations Centre for Human Settlements

COMMENTS: This is a quite straightforward microcomputer application that, in the case of the Housing Loans Scheme Division, is almost guaranteed to reduce the level of arrears on loans.

9. BUILDING AND VEHICLE-FLEET MAINTENANCE PROGRAMMES

POTENTIAL USERS: Administration and Finance Division; Local Government Division; and others.

SPECIAL HARDWARE: . None required.

SOFTWARE SOURCES: Simple non-commercial programmes are available for both building and vehicle fleet maintenance, or they can readily be developed by users familiar with BASIC or some other appropriate programming language. If sophisticated capabilities are required (particularly in the case of building maintenance), commercial microcomputer software is also available.

COMMENTS: These programmes can help maximize use (through efficient scheduling), minimize costs (e.g., through energy conservation) and prevent break-downs (e.g., by planning regular service programmes).

10. PROJECT MANAGEMENT

POTENTIAL USERS: Town and Country Planning; National Housing Department; New Villages Division; and others.

SPECIAL HARDWARE: None required, although complex projects are easiest to manage if a hard disk is available. (Hard disks should in turn be provided with a suitable backup system.)

SOFTWARE SOURCES: There are standard microcomputer programmes available for project management (like Harvard Total Project Manager, for example).

COMMENTS: Although the National Housing Department has a project management component built in to its QSS package, that Department, as well as other units in the Ministry, could benefit from a flexible, stand-alone package that offers sophisticated features (such as delay costing and optimum resource reallocation modules).

11. COMPUTER MODELLING FOR INFRASTRUCTURE PLANNING AND DESIGN

POTENTIAL USERS: Town and Country Planning Department; National Housing Department; Research Division; New Villages Division; and others.

SPECIAL HARDWARE: None required, although it may be desirable to enhance the basic equipment with a "mathematics co-processor" (such as the Intel 8087 or 80287) in order to reduce processing time.

SOFTWARE SOURCES: Microcomputer software is available for planning and designing housing programmes, road networks, street lighting, water-supply systems, sewerage systems, electrical power lines and so on. Software can be obtained from commercial sources (such as the Danish Hydraulics Institute) and non-commercial sources (such as the United Nations Development Programme, the United States AID Agency and the United States Department of Transportation).

COMMENTS: By virtue of their speed, computers make it feasible for Ministry staff to apply sophisticated mathematical techniques to the design and optimization of infrastructure systems.

12. SCIENTIFIC AND ENGINEERING ANALYSIS

POTENTIAL USERS: National Housing Department; Local Government Division; Research Division; and others.

SPECIAL HARDWARE: None required, although it may be desirable to enhance the basic equipment with a "mathematics co-processor" (such as the Intel 8087 or 80287) in order to reduce processing time.

SOFTWARE SOURCES: Microcomputer analysis can be applied to all sorts of building materials and structures, including soils, earthworks, posts, piles, foundations, beams, bridges and roads. Among commercial software products, TK!-Solver (for example) is a generic programme for processing mathematical formulas. Less suitable but still helpful for this purpose are standard spreadsheet packages, such as Lotus 123 or Supercalc. Non-commercial packages are available through universities, professional associations, etc. It is also possible for users with experience of BASIC or FORTRAN programming to write their own programmes for this purpose, if the appropriate languages are available.

COMMENTS: Microcomputers can speed up, standardize and encourage the use of sophisticated techniques for the numerous scientific and engineering calculations involved in all kinds of construction, including design specifications and stress analysis.

Therefore, the Ministry should probably continue as it has begun: that is, by providing most of its computer training in-house. As the number of microcomputers expands, it will probably be necessary to assign more resources to this function than at present. It may also be helpful to assign staff within each unit of the Ministry to provide training and support to microcomputer users and would-be users. In this respect two further points are worth bearing in mind:

- "Peer teaching" has been found particularly effective when it comes to microcomputer training. There is always a tendency to assume that teaching should be done by experts; however, with microcomputer training, people who have just learned to use a given programme can often do a good job of teaching others. This is because new users may be more keenly aware than experts of the precise learning process appropriate for a particular package. At the same time, teaching someone else provides an effective way for the new user to reinforce his/her understanding of the concepts and techniques of the programme.
- When it comes to identifying "peer" staff to assist with training and advice within each unit in the Ministry, it is often wise to wait and let appropriate people identify themselves by their actions. In any group of staff, it is likely that one or more will emerge as the local computer enthusiast — the person who is always willing to help others and the person whom others always seem to turn to. It is not so clear, however, who this person will turn out to be — whether it will be someone relatively senior or someone relatively junior, someone young or someone old, someone skilled or some-one unskilled. The reason is that computers just seem to appeal to certain people more than to others. For the sake of all concerned, it will be best for units in the Ministry if computer use is being promoted by someone who is genuinely enthusiastic about it. So, even if those staff members may sometimes neglect their own work for the sake of helping others, that may be small price to pay for productivity and job satisfaction in the unit as a whole.

However, training is not the only — or even the most important — factor that needs to be taken into account in developing computer literacy. There are at least two other critical factors besides training: motivation and access to equipment.

From the motivational point of view, it is essential that senior management be and be seen to be enthusiastic about the benefits and sympathetic to the difficulties of computerization. For most people, learning about computers amounts to a significant personal investment of time, energy and emotional commitment. It is unreasonable to expect staff members to make such an investment if they feel that the organizational commitment to computerization is ambiguous. Moreover, given the right motivation, a lot of learning can occur outside office hours: clearly, the Ministry benefits directly when this happens. So it makes good sense for the Ministry to encourage extracurricular computer use by (for example) providing access to computer equipment after office hours, encouraging a Ministry computer club or newsletter, subsidizing enrolment in evening courses and even assisting staff with the purchase of private microcomputers.

Finally, regular access to computer equipment is fundamental to the development of computer literacy. There is nothing intrinsically useful or enriching in learning about the effect of pressing certain keys on the keyboard when your computer is running a particular programme. It's rather like memorizing telephone numbers — useful while you are calling them regularly but soon forgotten when you do not. For most people, the only benefit that comes from knowing about computers comes from being able to practise and use that knowledge. Computer training for someone without regular access to a computer makes as much sense as driver training for someone who has no car. In both cases, the learning is much slower than it would be if more practice was possible; and, while the knowledge may one day "come in handy", that hardly provides a strong motivation. Anyway, the chances are high that the student will have to start all over again when he/she finally gets access to appropriate equipment.

In summary, it appears that the Ministry is generally proceeding sensibly if slowly towards computerization. Given the nature and scope of computer applications that are appropriate to an organization such as the Ministry, the current strategy of installing stand-alone microcomputers on a decentralized basis throughout the Ministry is probably the most cost-effective approach. Similarly, given the broad range of applications for which computers are intended in the Ministry, the policy of developing in-house resources for training and consultancy is also probably wise. However, the pace of both is slow. With confidence that its basic approach is sound, the Ministry should now, as an immediate priority, begin to increase the resources allocated to computerization. Accordingly, four principal recommendations are made:

1. Get additional equipment.
2. Build up the in-house support system for training and consultancy.
3. Start a computer-literacy support programme for Ministry staff.
4. Seek technical assistance for systems development.

It may be unpalatable but the fact is that there are probably small businesses in Kuala Lumpur with more computer power than the entire Ministry of Housing and Local Government! With a ratio of about one computer for every 300 employees in the Ministry, it is clearly impossible to contemplate widespread computerization on that basis. For computers to affect the operation of the Ministry in a fundamental way, the Ministry has to be ready to consider at least a five-fold increase in its microcomputer resources. Without a detailed study, it would not be prudent to recommend a precise equipment configuration for the Ministry; however, the following guidelines may prove helpful:

- Microcomputers come in various shapes and sizes. Unless there is a clear need for a "top-of-the-line" model, standard systems should consist of CPU, maximum memory, graphics capability, monochrome monitor, keyboard, two floppy disk drives and software for word processing, spreadsheet analysis and database management.
- At the present time, the wisest course is to buy equipment compatible with the IBM PC "standard". Brand-name as well as other equipment should be given equal consideration, with a view to determining the most advantageous source of supply.
- Unless there is a clear case for some other arrangement, it is recommended that microcomputers be distributed in workstations made up of two computers and at least one printer along with whatever other peripheral equipment is judged appropriate. Having two microcomputers at each site facilitates work scheduling, provides redundancy in the event of equipment failure and allows for better use of the peripheral equipment.
- Since there are steady improvements in the performance and steady reductions in the price of many peripherals (such as hard disk drives and other mass storage devices, colour monitors/printers, plotters, digitizers and optical character readers), these should be acquired in accordance with identified needs.

On this basis, a minimum configuration for the Ministry would consist of one workstation (i.e., two microcomputers and a printer) for each of the smaller units in the Ministry plus two each for the Local Government Division and the Administration and Finance Division, three for the National Housing Department and five for the Town and Country Planning Department. This comes to a total of 38 microcomputers and 19 printers.

It is worth emphasising how tangible some of the benefits of computerization in the Ministry can be expected to be. Take the case of the Housing Loans Scheme Division, for example, which currently administers its 2,800 loan accounts entirely by hand. Two-thirds of the accounts are said to be currently in arrears, to a combined extent of about $1 million. It is hard to think that an investment of $5,000-$10,000 in computer equipment would not soon be repaid by improved recovery of arrears.

Consider, too, the case of the Licensing Division, which sends letters and addresses envelopes to some 2,000 contractors every six months. If preparing and processing each mailing takes two person-months of clerical time alone, a computer which does the job in a few days soon pays for itself, quite apart from what it is used for in the other 50 weeks of the year. Or let a computer catch an error of only one one-hundredth of one per cent on a project of $25 million and the computer has paid for itself. Let a computer save half a centimetre in the specification of a building component in a 100-unit housing project, and that computer too has probably paid for itself.

These are all direct, tangible benefits. The indirect and intangible benefits of improvements in efficiency, public relations, job satisfaction, staff morale and so on may well be even more significant for the Ministry in the long run.

In short, a microcomputer is no more a luxury for the Ministry than a typewriter — and perhaps less of a luxury now that microcomputers can in fact replace typewriters for about the same price! The Ministry can continue to function without microcomputers, just as it could operate without any typewriters; but it would not be efficient for it to do so. Thus, in the long run, postponing computerization will not save money; on the contrary, it is likely to have the opposite effect.

Up to this point, the two-person Computer Unit in the Research Division is responsible for meeting the needs of computer users throughout the Ministry. Although this is probably a sensible approach and although the Unit seems to have done an effective job so far, it is clear that its resources will not be adequate in the face of the kind of growth in facilities proposed above. Accordingly, it is recommended that the size of the Computer Unit be doubled and that it be encouraged to establish a network of technical "contact persons" in each unit of the Ministry. These persons should be available to assist Computer Unit staff in identifying and responding to training needs in each unit; they should also act as the "front line" in dealing with users' problems.

The Computer Unit should also be prepared to assist departments and divisions in the Ministry to carry out systems analysis and systems design work. For complex requirements, it will be necessary to go outside the Ministry for assistance from private firms or experts. Here the Computer Unit can assist with advice and assistance in respect of selection of suppliers.

Finally, the Computer Unit should be equipped to provide advice and assistance in respect of additional hardware and software requirements. In order to keep abreast of such a fast-changing technology as that of microcomputers, it is essential that the Computer Unit be given the resources to do so. For example, the Unit should have a small budget for buying books, subscribing to magazines, testing new software, etc. Even though the Ministry may not make similar allowances for professionals in other fields, it is important to recognize that keeping up to date is a qualitatively different proposition for microcomputer specialists. The speed with which new products are being brought to market and the extent to which consumer experiences are published are quite possibly unprecedented. Ministry staff members will require assistance if they are to continue to provide their colleagues with the best and most up-to-date advice and assistance.

The Ministry will get the most out of its investment in equipment and support staff if it also takes the initiative in promoting computer literacy in the Ministry. After all, it takes time and effort to master a computer; the faster employees can be encouraged to do so, the sooner the Ministry will realize the benefits of its investment in equipment. There are numerous conventional promotional techniques that can be applied to microcomputers — speeches, seminars, awards, lotteries, etc. There are also several actions regarding microcomputers specifically that have proved successful in organizations elsewhere:

- facilitating staff access to office computers outside office hours (including weekends);
- establishing a staff computer club and permitting the use of office equipment for club purposes;
- publishing an informal, low-key newsletter containing hints and techniques tailored to Ministry applications and keeping users informed of how others in the Ministry are using their machines;
- reimbursing all or part of the cost of approved computer courses that employees attend in their own time; and
- assisting employees to buy their own microcomputers, either by using the Ministry's purchasing power to secure favourable prices for employees or by directly subsidizing (e.g., by an interest-free loan) the cost of the equipment.

One of the key aspects of this case has been the number and diversity of potential computer applications in the Ministry. It has also been argued, sometimes explicitly and sometimes by implication, that the hardware and software necessary to realize these potential applications is relatively inexpensive. Thirdly, it has been suggested that the Ministry should continue to proceed as far as possible on the strength of its own resources.

Nevertheless, it would obviously be helpful for the Ministry to have the services of an expert adviser for six months or a year. Such an expert should be not just a computer specialist: he or she should be experienced specifically with microcomputers and their application to the planning and management of human settlements. The expert should also ideally be someone who is good at teaching non-specialists about computers and someone who is sensitive to the impact of computers on organizations. For the expert would in fact have a dual role. On the one hand, he or she would work with the staff members of the Computer Unit in order to strengthen their capacities and enable them to assume his or her functions; on the other hand, the expert would work directly with staff members in other units, assisting with systems analysis and system design in accordance with their particular requirements.

Case 8

Management applications (Thailand)

The Land Transport Department of the Ministry of Communications was established under the Land Transport Act of 1979 (BE 2522). Its primary responsibility is to license, regulate and promote the safe and efficient vehicular movement of passengers and goods in Thailand. The effect of the Act is to create two classes of vehicles, operators, employees and facilities: those which come under the Act and those which do not. Essentially, the Act applies to all civilian vehicular transportation except that provided by private passenger cars (including taxis) that do not carry more than seven passengers, motor-tricycles (*tuk-tuk*), motorcycles and tractors, all of which are licensed and regulated by the police under the Motor Vehicle Act and the Traffic Act. However, this distinction is to be removed in due course and responsibility for vehicles now under police jurisdiction transferred to the Land Transport Department.

The Department is organized into six Divisions, one Sub-Division and 72 district offices, one in each province (or *changwat*). Their duties regarding operators, vehicles and personnel coming under the Land Transport Act are as follows:

- Transport Regulation Division — to license operators of transport services, including the type of service, routes, fares/tariffs and number of vehicles; and to build bus shelters in Greater Bangkok and bus terminals throughout the country.
- Registration and Taxation Sub-Division — to maintain the Central Vehicle Registration Record (CVRR) and to register and tax all vehicles.
- Transport Safety Division — to issue licences to drivers, conductors, fare collectors and passenger-service staff; and to promote safety through training, testing, certification of private schools and collection of accident statistics.
- Transport Supervision Division — to regulate and inspect operators of transport services (including investigation of complaints,

regulation of hours of work and provision of advice on accounting and other business practices) and to maintain performance records of drivers and conductors.

- Transport Engineering Division — to regulate and inspect vehicles (including construction of inspection centres), to maintain vehicle history records and to carry out research on matters related thereto.
- Technical and Planning Division — to carry out research and analysis on all aspects of road transport, with a view to establishing optimal routes, schedules, fares/tariffs, licence allocations and user charges.
- Office of the Secretary — to provide administrative services for the Department (including personnel, accounting, payroll and supply) and to collect and account for revenue from licences and other fees.

The structure and function of district offices are modelled on those of the central administration. Their powers and duties extend essentially to licensing and regulating those activities that are confined to their respective jurisdictions (the *changwat*). Thus, operator licences for private use or for routes/hire within a *changwat* are normally processed at the district office (subject to approval of a local Land Transport Control Board), while licences for services to and from Bangkok or another *changwat* are referred to the central administration (subject to approval of a national Land Transport Control Board). Similarly, vehicles are normally registered and taxed by the district office where their operator is based. Finally, a district office may also issue and renew personnel licences for people domiciled in its *changwat*.

The operations of a typical district office are shown in Table 7. Lamphun is a *changwat* immediately to the south of Chiang Mai (the second largest city in Thailand), about 700 km north of Bangkok. The total population of Lamphun *changwat* is about 300,000; as of mid-1985, it had 834 operators, 1,868 vehicles and about 12,000 personnel licensed under the Land Transport Act; the local office has a staff of nine. According to police records, there were 42,607 other licensed vehicles in Lamphun at the beginning of 1984. Compared to the other 71 district offices, Lamphun reportedly stands about fortieth or fiftieth in terms of its volume of "business"; however, its pattern of activity is probably typical of most district offices. As for operator licences, their relatively long period of validity (five or seven years) means that district offices have to cope with a low and fairly steady volume of business each month. Because vehicles are taxed annually (although payment may be quarterly), vehicle processing tends to follow a quarterly cycle, especially in the last calendar quarter. In the two months of December

Table 7 Summary of Land Transport Office transactions in the *changwat* of Lamphun for the first nine months of fiscal year 1984–5 (Thailand)

	Oct	Nov	*1984* *Dec*	1985 *Jan*	*Feb*	*Mar*	*Apr*	*May*	*Jun*	*Yr to Date*
Operator Licence Approvals (of Total 834):										
All Categories	10	15	18	15	14	10	11	18	12	123
Vehicle Inspections (of Total 1,868):										
New Vehicles	11	10	6	11	9	7	9	9	9	81
Old Vehicles	64	143	602	183	53	166	110	67	155	1,543
Total	75	153	608	194	62	173	119	76	164	1,624
As % of Total	4	8	33	10	3	9	6	4	9	86
Vehicle Registration and Tax Collection (of Total 1,868):										
New Vehicles	11	10	6	11	9	7	9	9	9	81
Renewals	64	123	598	181	55	162	77	20	153	1,433
Transfers	31	46	49	47	35	55	80	83	37	463
Total	106	179	653	239	99	224	166	112	199	1,977
As % of Total	6	10	35	13	5	12	9	6	11	107
Personnel Licensing (of Total 12,000):										
Drivers (new)	61	91	55	78	61	69	29	80	83	607
Drivers (renew)	359	291	274	298	262	363	291	279	330	2,747
Others	2	7	3	13	4	6	2	1	3	41
Total	422	389	332	389	327	438	322	360	416	3,395
As % of Total	4	3	3	3	3	4	3	3	3	29

Source: Land Transport Office, *Changwat* of Lamphun (August 1985). Data in parentheses are estimates of total population in the *changwat*. Most operator licences (828 of 834) are for private use (classes 40 and 80); operator licences are good for five or seven years (depending on the class). All vehicles must be inspected at least once a year when the road tax is paid. On approval, vehicle inspections can be done in any *changwat* office, but registration and taxation must be done in the *changwat* where the operator is based. Data on registration and taxation do not include notifiable changes not requiring inspection (such as repainting). Personnel licences are renewable annually (or every two years on request), most of them in Bangkok.

and January, a district office has to inspect and collect tax from nearly half of all the vehicles in its jurisdiction. Additional inspections and processing are required after accidents, repairs, etc. About one-third of the personnel licences are renewed locally, and this represents a fairly steady workload for the district office over the course of the year.

From these functions, it is possible to identify three broad areas where data management plays a key role in the activities of the Department: records management, research and planning, and departmental administration.

1. Records management

The Department maintains five main sets of records pertaining to operators, vehicles and personnel coming under the jurisdiction of the Land Transport Act (LTA):

- a Transport Operators Roster, including details of the routes and number of vehicles licensed to each operator;
- a Central Vehicle Registry Record, including details of the registration and taxation status of every licensed vehicle;
- a Transport Personnel Licence File covering all licensed drivers, conductors, fare collectors and other passenger-service staff;
- a Driver/Conductor Performance History Record containing details of the accident and employment history of each licensee; and
- a Vehicle History File, including details of inspections and accidents to all licensed vehicles.

The principal source of data for at least four of these five databases is the offices through which the Department deals with the public (i.e., in Bangkok and the 72 *changwat*). The approximate scale of the databases is illustrated in Table 8.

2. Research and planning

The main undertaking to date has been an annual survey of inter-city goods-movement (origin-and-destination) involving some 600,000-700,000 records of about 60 characters each. Data-processing has up to now been done for the Department by the National Statistical Office; however, the survey will reportedly not become a regular event.

There are several other areas of research and planning where computer applications are possible. Among these are: statistical analysis (e.g., of accidents), linear programming (e.g., for route selection), financial analysis (e.g., for setting fares and tariffs), forecasting (e.g., for population projections) and modelling (e.g., for transport demand forecasting).

Table 8 Summary of overall departmental records management requirements (Thailand)

Licensed Transport Operators (1984)

Class	*Type of Service*	*Licensed Range*	*Numbers of Licensed* *Operators*	*Vehicles*	*Routes*
10	Bus (fixed-route)	Bangkok Metro Area	9	9,316	298
		BMA-Provinces	1	3,409	160
		Provincial	203	7,441	354
30	Bus (unrestricted)	Bangkok Metro Area	264	1,772	
		Provinces (36)	114	537	
		Other Provinces	-	-	
40	Bus (private)	Bangkok City	2,808	3,891	
		Rest of Country	-	-	
70	Truck (for-hire)	Bangkok City	213	8,154	
		Rest of Country	-	-	
80	Truck (private)	Bangkok City	7,722	28,750	
		Rest of Country	-	-	

Licensed Transport Vehicles (1984)

	Estimated Number of Vehicles Under *Land Transport Act*	*Police Control*
Bangkok Metropolitan Area (BMA)	60,000	1,090,000
Rest of the Country	240,000	2,910,000
Total	300,000	4,000,000

Licensed Transport Personnel (1985)

	Estimated Number of Personnel Under *Land Transport Act*	*Police Control*
Bangkok Metropolitan Area (BMA)	74,914	-
Rest of the Country	389,754	-
Total	464,668	900,000

Source: interviews with officials in the Transport Regulation Division, Land Transport Department, and police officers; a dash means data are not available. Note that, especially in the case of personnel licences, official data probably underestimate the actual numbers.

Most of this work is done by the central administration, although there may be local applications in terms of workload planning, local operator-licence approvals, bus-terminal administration, driver education and training (including video simulations for testing co-ordination and reaction-time), etc.

3. Departmental administration

No computer applications have been undertaken for administration of the Department itself but there is scope for it in several areas: for example, personnel and payroll (the Department has about 1,500 full-time staff divided more or less equally between central and district administration); budgeting and inventory; and — perhaps most important of all — accounting for the approximately THB 750 million collected in taxes and other fees each year by the Department in Bangkok and through its 72 district offices.

Most of these applications (including the records management functions described above) apply equally to the central and district administrations.

Faced with these requirements, the Department has begun to prepare for computerization. In 1983, an Elbit/KeyPact 4800 data-entry system was acquired from Israel, consisting of 16 terminals, a 4811 CPU, an eight-inch floppy-disk drive, a disk cartridge unit, a CDC 1860-5 magnetic tape unit and a CDC 1827-7 line printer. With this equipment, the Department has begun transferring some of its vehicle and personnel records to machine-readable form (magnetic tape). Two files are in the process of being created:

- a Vehicle Registration and Taxation File containing data (631 characters) in five records on each LTA-licensed vehicle, of which about 50,000 (about 17%) have so far been transferred to tape; and
- a Transport Personnel Licence File containing data (596 characters) in three records on each person licensed under the LTA, of which about 80,000 (again, about 17%) have so far been transferred to tape.

Thus, the Land Transport Department faces a critically important decision. Not only is the Department proposing to automate a considerable proportion of its activities — which is a daunting enough task on its own, but, at the same time, the Department faces the added responsibility of three million vehicles and nearly one million licensees which currently fall under police jurisdiction. In terms of numbers, this means adding about 10 times the number of vehicles and twice the number of licensees that the Department now handles — to say nothing of the problem of unlicensed vehicles and drivers, which appears to be more serious in relative as well as absolute terms in respect of police-regulated vehicles and drivers than it is for those currently regulated by the Department.

Computers clearly hold the key to coping with such vast quantities of data. Yet, computerization will almost certainly bring problems and opportunities of its own to an already challenging situation. The potential for improving performance and productivity through computerization is clearly there, but a lot is going to depend on making sound decisions about how to proceed in the next few months and years.

The main feature of the Department's vehicle and personnel databases is that they are already big and could get a lot bigger — in excess of a gigabyte! — with the transfer of responsibilities from the police. Computers can handle large databases, of course; but, even with computers, the larger the database the more it costs to operate — more in terms of initial capital cost for the equipment, the time required to process the data, the number of personnel and terminals needed for data-entry and so on. Even at the "blinding" speed at which computers operate, it still takes longer to process a file of 100 megabytes than it does a file of only 10 megabytes. Thus, the most difficult questions in changing from manual to computer-based systems of record-keeping often involve deciding what data to leave out rather than what data to put in. (It is worth noting that, using its 16 terminals, the Department has taken two years to transfer about one-sixth of its current databases to magnetic tape. If that trend continues, it will take five more years to complete the task — and this makes no allowance for growth in the existing databases or for entering data from the vehicle and personnel records to be transferred from police jurisdiction!)

For one thing, there is a tendency to forget that manual systems, like computer-based systems, tend to reflect the technology they use. So, a system that works well manually — one that may even be the fruit of years of experience and many refinements — may not be the optimal design for a computer-based system. For example, a good manual system may include built-in devices for facilitating cross-referencing: the same data recorded in several different places or an internal record number to facilitate searching and sorting. Neither of these is necessary in a computer-based system, since the computer can as easily search one part of a record as another or order names as easily as numbers.

For another thing, not all data are equally desirable for storage in a computer. As a general rule, data which do not need to be processed — that is, do not need to be edited, searched, sorted, added up, averaged, correlated, graphed or otherwise analysed and manipulated — do not really need to be computerized. In fact, such data may often be accessed more easily and more conveniently in ordinary documentary form than by computer. For example, as long as there is no need to find or identify a person on the basis of the names of his parents, there is little point in putting such data in a computer file; they could as well be recorded on paper in a file in a filing cabinet (or on microfiche, if storage space is a factor). Similarly, accident data can be divided into data which can usefully be processed — date, time of day, location, number of injuries/fatalities, extent of property damage, etc. — and data which are unlikely to warrant processing — such as descriptive information about the nature and causes of the accident. The case for putting the first kind of data into a computer is clearly stronger than the case for the second.

This is no reflection of the intrinsic importance of the data themselves, only of the way in which the data are to be used. It is well to remember that computers are primarily tools for data-*processing*, not data *archiving*.

Thus, computerization is not an all-or-nothing, go-for-broke type of proposition. Computerization does not require that, once committed, you turn everything over to the computer. Instead, computerization should be seen as complementary to manual processing. Certain kinds of data management can undoubtedly be done more quickly and accurately on a computer than they can manually, but the best way to take advantage of the power of the computer does not lie in assuming it has an unlimited capacity for data storage — that all existing data can and should be computerized. Thus, even if you have a manual database management system that is working smoothly, computerizing the manual system requires renewed analysis and planning — analysis of data availability and planning of information needs — of what you've got and what you want to do with it.

Prior to computerization, therefore, the Land Transport Department should be sure to take a close look at the contents of and reasons for each of its databases, and to carry out a formal systems analysis of data sources and information needs in relation to computerized data-processing. It is worth underlining the fact that when you have a database of 3 million records, every single character in a record adds three megabytes to the overall size of the database. Add a whole new field of (say) 10 characters and you add 30 megabytes to the size of the file. Thus, for all their advantages, computers impose a degree of discipline on the organization and structure of a database that is much greater than that of a manual system. In a computer-based system, every byte counts; so every character that goes into a computerized database needs justification and every field should be intended for specific kinds of data-processing.

For example, a thorough departmental systems analysis would examine the organization and structure of each proposed computer database to see how far:

- field sizes could be reduced through abbreviating, truncating, coding or other techniques;
- record sizes could be reduced through omitting fields that are redundant and fields that are unlikely to figure in data-processing (searching, sorting, analysing, etc.)
- file sizes could be reduced through fragmentation based on data-entry and search strategies planned to meet anticipated user needs.

These are not absolute rules, of course; it is possible to become fanatical about database design to the point where ease of use is sacrificed to

Table 9 Preliminary summary of factors involved in systems analysis of departmental computer database requirements (Thailand)

	Transport Operators Roster	*Central Vehicle Record Registry*	*Transport Personnel Licensing File*	*Driver & Conductor History File*	*Vehicle History File*
Est. file size	1 Mb	200 Mb	300 Mb	Data not available	Data not available
Est. police files	None	900 Mb	200 Mb	Data not available	Data not available
Adding new records	Occasional only	Frequent; growth estimated 10% p.a.	Frequent	Frequent	Frequent; growth estimated 10% p.a.
Editing existing records	Occasional only, as licences good 5–7 years	Very frequent, as records updated at least annually, sometimes several times per year	Frequent; licence renewed annually or every two years; also occasional changes in biodata	Fairly frequent, as operators and police report accidents	Very frequent, as annual inspections plus occasional other mechanical alterations
Sorting	Yes, depending on analytical requirements	No; file maintained in order of registration plate number	No; file maintained in order of province and licence number	No; file maintained in order of province and licence number	No; file maintained in order of registration plate number
Searching	Frequent on various fields	Frequent on key-field; rare on others	Frequent on various fields	Frequent on key-field; rare on others	Frequent on key-field; rare on others
Sub-string search	Probably useful	Unlikely useful	Unlikely useful	Probably useful	Probably useful
Specialized hardcopy output	Operators Certificate	Owners Certificate and Tax Sticker	Nine different personnel licences	None	None
Mathematical processing	Route planning	Fleet planning and operating costs	Statistical analysis only	Accident analysis	Accident analysis
Potential for file disaggregation	File too small	Could be done on basis of class of vehicle licence	Could be done on basis of issuing province or class of licence (9)	Could be done on basis of issuing province or class of licence (9)	Could be done on basis of class of vehicle licence
Remote input/output	Not required	Highly desirable at licensing office	Highly desirable at licensing office	Not required	Highly desirable at inspection station
Data transfer requirements	Not required	Police and/or Highway Patrol	Police and/or Highway Patrol	Not required	Not required

Source: interviews with various departmental officials (August 1985).

elegant design. Too much abbreviating and coding can become counter-productive if it makes data entry inefficient and data interpretation obscure. Similarly, you may regret omitting certain data when you subsequently find you would like to use them. So good judgement is needed for good database design. However, the basic principle is that the presence of any data in a computerized database needs justification in terms of what the computer is going to do with or to them. It is not enough that data have to be stored somewhere; so put them in the computer. Table 9 gives a preliminary listing of some of the factors that should be taken into account in carrying out a thorough analysis of data and information requirements in the Department.

The main point that needs to be made in connection with system design is that data destined to go into a computer should be put there as soon as possible. In other words, data for a computerized database should be put into electronic form as close to the point where they are collected as possible. This speeds data transfer, makes data-processing easy and fast, and reduces the likelihood of error.

Although this principle of getting data into electronic form as early as possible may seem self-evident, there is a tendency for organizations setting up computer systems for the first time to work on the opposite principle. That is, the computer system is designed so that records are processed manually (just as they always have been) and then, as a final step in the process, they are sent to the computer department for entry into the computer. The effect of this procedure is to deprive the organization of many of the benefits they expect to get from computerization. The effect is to exclude the computer from much of the operational work of the organization and to reserve it for (often undefined) research, policy and planning uses.

All of this applies particularly to the Land Transport Department, which has large data flows to provide for and extensive operational data-processing requirements to meet. Almost all of the data in the departmental databases are collected as part of the day-to-day operations of the Department. Data collection is not a discrete activity by a small section of the Department; data collection lies at the heart of the Department's principal responsibilities of regulating and supervising land transport under the Land Transport Act. Moreover, most of the data are collected at remote sites (public offices in Bangkok and *changwat* offices) and have to be transferred to various other locations in the Department. In circumstances such as these, data that are to go into a computer should do so as early as possible, preferably at the point where they are first collected.

This is particularly true if, as in the case of the Land Transport Department, collectors of data are also information users. Early computerization of data at the local level will not only help with the central administration of records; it will also do a lot to help local and *changwat* offices manage their own affairs efficiently and encourage them to provide data to the central computer on a timely and reliable basis.

There are two basic approaches to providing this kind of computer capability. One approach is to set up a central "host" computer and then provide on-line access to it from remote sites. Access from the remote sites may be over special cable or (if they are available and suitable) dedicated telephone lines. Data transfer between the host computer and the remote station occurs in "real time"; in other words, the remote station functions effectively as a long-distance terminal to the host computer. However, remote sites may also have some independent (stand-alone) capacity of their own.

An alternative to decentralized access to a central computer is distributed processing. Essentially, this involves setting up independent (stand-alone) computers and then linking them through a network. This network may take several forms. At its simplest, the network may be no more than a procedure for exchanging (even mailing) disks and tapes containing data from one computer to another; or the network may mean periodic communication between computers via modem and standard telephone lines; or the network may have a permanent and tangible nature in the form of dedicated lines or cables. The computers involved in the network need not all be the same nor have the same capacities, although some minimal compatibility will be important, depending on the kind of network to be used and the data to be transferred.

Briefly, the advantages of the first are speed and efficiency, while the advantages of the second are low-cost and redundancy. Where large quantities of data have to be transferred quickly and frequently, the greater cost and vulnerability of the centralized system may be justified. Where data files can fit conveniently on disks and tapes and where the speed and frequency of transfer are not critical, it is usually wiser to settle for the simpler procedure of providing the required computer capacity at remote sites and arranging to "up-load" and "down-load" data to and from a central computer as and when required, using whatever technology (physical transfer of medium, telephone link or dedicated network) as may be appropriate in each case.

For the Land Transport Department, the case for distributed processing would seem to be strong. For most departmental data, the speed of data transfer is not critical: if data from a *changwat* take days to reach Bangkok instead of seconds, it does not matter a great deal. If transfers are effected only several times a month instead of several times a day, the work of the Department can still proceed smoothly and efficiently. Moreover, distributed processing would provide local and *changwat* offices with a valuable data management resource for their own record-keeping, administration and planning.

The theme of this case can be reduced to one main recommendation:

> that the Department should make a thorough analysis of its data sources and information needs as a basis for designing an overall computer policy for the Department.

Several key parameters for such an analysis have already been suggested in what has been said so far:

- the need to provide the maximum of information from the minimum of data;
- the need to provide computer access as close to the point of data collection and information use as possible; and
- the need to plan explicitly for the transfer of police responsibilities for vehicle and driver licensing to the Department.

Once this analysis is completed and a comprehensive plan for the design of a computer system for the Department is available, there should be a phased programme of implementation. This programme should be based not so much on functional logic — data-entry, then data-processing, then data output — as on the need to co-ordinate equipment, training and information needs as much as possible so that they reinforce one another.

For example, if equipment is to go first to some of the *changwat* offices (as this case has proposed), that decision should be supported by an emphasis on operational applications, provision of simple but wide-spread user training and availability of facilities for communications with a central computer — even if it is not (yet) much larger than any of the *changwat* computers. On the other hand, if equipment is to go first to the Divisions of the central administration in Bangkok, that decision should be supported by an emphasis on planning at the expense of operational applications, provision of selective and professional types of training and availability of facilities for printing, mapping and graphing results. Whatever strategy is chosen for implementing distributed processing should in turn be co-ordinated with the timing and priorities associated with procuring a central computer, so that this too reinforces and is reinforced by other activities that are going on. The key to successful implementation of any computer system lies in a well co-ordinated and well thought-out plan of attack.

Part three

Information systems

Part three

Information systems

If there is an "El Dorado" in computerization, it is probably information systems. The one thing that can be counted on to win converts to computers, especially at senior-management levels, is the vision of being able to push a button and get instant answers to vital questions. The reality, however, often falls short of the vision.

The main reason for the shortcomings of information systems is almost invariably a lack of definition of purpose at the outset. This stems in turn from a tendency to believe that the justification for information systems is self-evident. We live in an age of information. We take it for granted that the more information we have, the better position we are in — or, at least, that more information never does any harm and almost always does some good. Like motherhood, information systems tend to be regarded as intrinsically virtuous.

In practice, however, information is rarely more than a means to an end. Information systems are largely utilitarian. If they contribute to planning and decision-making, well and good; if not, they have little purpose. Thus, the criteria for designing an information system are not inherent in the information but rather in the environment whence the information is to be derived and in which it is to be used.

This means that the design of an information system depends essentially on the purpose it is meant to serve. There is no fixed quantity or even kind of data that must, of necessity, go into a given information system. There is nothing in the data themselves that makes them a necessary part of one information system but not of another. The subjects included in an information system, the level of detail it provides and the historical period it covers are all matters of choice. These choices depend primarily on external factors, such as (a) the availability of data to go into the system and (b) the kind of uses which the system is meant to serve. Clearly, there is little point in creating an information system based on data that are either unavailable or (if available) irrelevant.

There are two important factors in designing information systems. First, an information system has to provide better information more quickly and efficiently than alternative sources. Secondly, an information system has to present data in ways that make it easy for people to use the data. If an information system does not meet these criteria, it will not be used.

The key to providing good information is to design the information system around data which are readily available and to build into the system a regular and continuous process of keeping the data up to date. If possible, information systems should be based on data that are already being collected (or, for relatively little extra cost, could be collected) as part of some agency's normal, everyday operations. In that way, there can be some assurance of a regular supply of up-to-date data. This may mean relying on untraditional sources of data, such as administrative or financial records, instead of formal census or survey data. However, the alternative of relying on data specially collected for the purpose is to render the information system perpetually hostage to a continued willingness to make that special collection of data — which may also leave the information system seriously exposed in the event of financial cutbacks.

How quickly information can be made available depends largely on how accessible it can be made relative to alternative sources. If it merely regurgitates census data, for example, an information system will have to be very accessible in order to persuade people to use it instead of the census report sitting on their desks. Access speed also depends on size. Even with computers, large quantities of data take longer to process than small quantities. While it may not seem much to want to add one more field to a large database, the effect of larger size is almost always to make processing slower. Information systems are not archives to be used for storing data "for the record" just in case someone wants to look it up. Information systems are tools whose value is related to their usefulness for specific purposes. As a general rule, data should always have to justify their presence in an information system in terms of their usefulness for these purposes.

The other key element in the design of an information system is what it enables users to do with its data. A good information system has to do more than just present data for users to look at; data have to be presented in a form and in a context that facilitate their use for planning and decision-making. In practical terms, this means choosing data and creating a structure for the information system that allow users to compare, manipulate and analyse data; to examine assumptions, test hypotheses and set criteria of evaluation; and to summarize results and present findings in an efficient and effective way. Information systems are more than just data banks; they are data banks with a purpose. An information system is effective to the extent that it facilitates this purpose. Indeed, the most difficult kind of information system to design is one specified only as for "general" use, for this gives little guidance as to what uses the information system should be designed to support.

In all these respects — keeping information up to date, processing it quickly and presenting it flexibly — computers provide an ideal environment for information systems. Data which need to be updated frequently, processed extensively or presented in a variety of formats are ideally stored on a computer. By the same token, data which do not often change, do not often need processing and are not often reproduced in reports or other documents have little place in computerized information systems. Given the new access to computer power offered by low-cost, user-friendly microcomputers, computerized information systems have become a viable prospect for many human settlements agencies.

This section presents four cases which illustrate in different ways the challenge of designing effective information systems in the field of human settlements. The first case (from Indonesia) describes the difficulties of trying to build an information system around "available data" when availability is in fact their chief merit. By way of contrast, the second case (from India) shows the problems of designing an information system "from the top down", with little or no regard for the supply of available data or the demands of likely users. The third case (from the Philippines) describes the successful development of a human settlements data bank but contrasts this with the need for an information system that is operationally useful for planners and policy-makers. The fourth case (from Mauritius) presents the design of an information system for managing publicly owned land, in which the regular supply of data and the likely needs of users are specifically taken into account.

In the end, the cases illustrate a number of key lessons about the design of information systems.

1. It is almost always best to proceed in stages. For one thing, it may not be possible to anticipate how an information system will be used until there is some experience with it. Secondly, it usually makes sense to get a limited system in operation as soon as possible. Not only will this help win political support for the large project; it may even help pay for it, if the initial effort is focused on something like property tax or rents from public housing.
2. Get data into electronic form as soon as possible. This speeds processing, facilitates data transfer and reduces the possibility of transcription errors.

3. Decentralize computer access to the maximum extent possible. Even if it means making do with "basic" equipment (ordinary PCs instead of ATs, for example), the more people who can benefit from and contribute to an information system the better.

Case 9: Data aren't everything (Indonesia)

In Indonesia, the Directorate of Urban and Regional Planning, which is part of the Directorate-General of Human Settlements (*Cipta Karya*) in the Department of Public Works, is the agency responsible for implementation of the National Urban Development Strategy (NUDS). One of the components of the NUDS was to be an appropriate information system. By early 1985, two steps had been taken towards creation of the proposed information system: a VAX 11/730 had been supplied and installed, and extensive data had been collected in the course of various project studies. However, it soon became apparent that there was more to creating an information system than just putting data into a computer.

Clearly, the main impetus for the NUDS information system came not from the demand side but from the supply side — from the existence of what is arguably the most extensive human settlements database ever assembled in Indonesia. However, the usefulness of an information system depends more on the quality than the quantity of its data. In particular, it is important for data in an information system to be relevant to the needs of its users, as up to date as possible and accessible to users in a form that is convenient. For this reason, it is better for the design of an information system to compromise its scope rather than its relevance, its timeliness or its accessibility.

Thus, the next step for NUDS should be to design and implement a preliminary, "quick and dirty" information system. First, this would involve identifying a core database chosen for its likely relevance and ease of regular updating. Secondly, there should be an effort to make the VAX minicomputer as accessible as possible to potential users, by facilitating communication with microcomputers and terminals throughout the Ministry, by providing user-friendly software interfaces on the VAX and by providing adequate computer training and support for users.

Case 10: Top-down versus bottom-up design (India)

In India, the Town and Country Planning Office (TCPO) of the Ministry of Works and Housing has taken a leading role in promoting the development of information systems for human settlements planning. Since as early as 1977, TCPO has been helping to organize meetings and workshops on the subject. In 1981–2, TCPO succeeded in persuading the state governments concerned to collaborate on two pilot studies, one in Chengalpattu (Tamil Nadu) and one in Anand (Gujarat). At the same time, TCPO has been working on a number of in-house "prototype" information systems of its own.

The most striking conclusion about the TCPO experience is that, in spite of having received far less input in terms of manpower and computer resources, the more modest "prototype" systems are closer to realization than the more ambitious pilot studies. The reason is that the former have gone further in recognizing the dynamic interrelationship between the data, the users and the capacities of information systems. In particular, it is a mistake to adopt a rigidly linear approach to system design, in which the first step is to decide what data are to go into the system, the second step to decide how the data are to be processed and the last step to identify and deliver the system to its users. It is wiser to work on all three aspects of an information system at once — data, users and processing capabilities — refining the definition of first one and then the other within the constraints of all three, until there is a design that appears to balance the needs and resources of all.

In short, development of a comprehensive urban and regional information system (as envisaged in the pilot studies) is an ambitious undertaking. For this reason, it should be approached on an incremental or prototype basis. Rather than think in terms of a single information system, it would be wise to envisage a "system of systems" — a series of functioning prototype information systems, each serving an immediate practical need but each designed so that it could eventually be linked into a large and comprehensive information system. Not the least advantage of such a strategy is that it provides immediate results, which can serve as a useful fall-back in the event that the larger goal proves elusive.

Case 11: Data banks and information systems (Philippines)

In the Philippines, the Ministry of Human Settlements is the chief government agency responsible for land-use planning and zoning. Since the mid-1970s, one of the goals of the Ministry has been to develop and maintain a central computerized Planners' Data Bank.

The Data Bank has three main components: a statistical database, a geographical database and a project-management database. Four series of publications have appeared based on the Data Bank: settlement profiles, atlases, background reports on town planning and service-delivery profiles. In addition there are plans to extend data-gathering and reporting down to the level of the *barangay* (of which there are about 42,000 in the country as a whole), specifically in respect of community profiles, various development indicators and data on standards of service in respect of "basic needs" throughout the country.

This case provides an assessment of the Planners' Data Bank and its achievements over the past decade. In general, the Data Bank is not as flexible or as accessible as it should be, if it is to serve as a tool for planners. Use of the Data Bank is limited to production of hard-copy reports; there is no interactive use of the Data Bank. Moreover, some of the data contained in the Bank are of doubtful relevance, timeliness and even accuracy. Accordingly, the case recommends development of a decentralized and diversified information system, one aimed at meeting a more explicit set of planning and management objectives. Such a Human Settlements Information System (HSIS) would naturally have to be supported by sufficient equipment, programming, training and support to make it genuinely accessible to staff in Manila and at the district-office level.

Case 12: A land management information system (Mauritius)

In Mauritius, the Ministry of Housing, Lands and the Environment is the governmental agency responsible for the administration of Crown Lands and Leases. The Ministry has just obtained an IBM PC AT microcomputer and proposes to use it for managing its land records. At the present time, there are approximately 10,000 Crown Land lots, of which about 7,000 are subject to lease or vestment involving a total of some 12,000 leases or other legal documents. Naturally, the Ministry needs to maintain and monitor records on the status of every one of these lots and leases.

Land records are currently kept entirely by hand using two separate databases, the *Domaine Book* which records Crown Land ownership, and the leases themselves, which are accessed through an *Index of Leases*. There are also records of transactions in Crown Lands and of payments on lease accounts. There are three main reasons for computerizing these records as quickly as possible: security against loss or damage; fast checking, processing and analysis of records; and improved administration (including prompt collection of rents and renewal of leases when due).

Thus, the case provides a detailed design for a land-management information system geared to the Ministry's existing computer capacity yet designed for expansion if additional resources become available. The case also discusses how this information system could in due course be integrated with other data management systems for use within the

Ministry (e.g., for surveying, financial analysis and town planning) and in other parts of the government (e.g., in the Registrar-General's Department, which is the official repository for all governmental records). As such, the proposed system meets two of the key criteria for an effective information system: it fills an immediate operational need and it represents a useful incremental step towards eventual development of comprehensive planning information systems.

Case 9

Data aren't everything (Indonesia)

The Department of Public Works in the Directorate of Urban and Regional Planning (*Tata Kota dan Tata Daerah*) of the Directorate-General of Human Settlements (*Cipta Karya*) is setting up a computerized information system on human settlements as part of a United Nations-assisted project (INS/78/059) to design a National Urban Development Strategy (NUDS).

According to the original project document, the overall goal of the NUDS project is:

> to obtain a more balanced development through a more appropriate and equitable allocation of development resources to meet better spatial distribution of population in line with the comprehensive objectives of the State Development Policy Guidelines (GBHN) [and] by assisting and strengthening the institutions charged with development policies.

This overall goal has subsequently been translated into a number of more specific "operating themes" and "operating objectives". According to an internal project document in June 1983, the four "operating themes" are:

- "a realistic understanding of the social and economic forces that shape urban development";
- a commitment to providing recommendations "that will meet the practical needs of government decision-makers";
- "continuing interaction with government departments" during the life of the project; and
- distributional equity throughout the country.

These themes are meant to provide the context or framework for achievement of three fundamental "operating objectives" that have been defined for the project, namely, improved spatial development policies; an integrated urban development strategy; and an institutional framework capable of updating and implementing this strategy. One explicit element in the third of these objectives is "development of an information system to respond to the need for regular updating and monitoring of strategic plans".

The significance of the NUDS project was underlined in an official report issued the following year. In March 1984, UNCHS (Habitat) sent a Mission (Kenneth Watts) to Jakarta to assess the contribution of United Nations assistance to human settlements planning and management in Indonesia over the previous five years. The Watts report concluded that, while remarkable progress had been made in upgrading basic urban services in Indonesia over the past 15 years, it was doubtful how much of this could actually be attributed to good planning. For one thing, the report noted that comprehensive development studies had "seldom led directly to implementation of (urban) projects and programmes". Secondly, the report pointed out that the government and the international banks were still resolutely sectoral in their approach to implementation. Thirdly, the report concluded that, "over and above everything else, the work so far lacks a sense of priority". Although the case for improving urban infrastructure is "self-evident to a degree", there is little evidence in Indonesia that the contribution to overall urban development of all the investment in urban infrastructure to date has been much more than the sum of its parts.

Essentially, it was these basic weaknesses in urban development strategy hitherto — lack of integration in project design and lack of co-ordination in project implementation — that the NUDS project was intended to address. The output of the NUDS project is to be nothing less than a complete integrated urban development strategy. The Watts report went on to summarize the NUDS approach in the following terms:

> First, identify the "functional potential" of each major city and town in Indonesia.
>
> Secondly, design "packages" of government programmes appropriate for cities and towns with different levels of "potential".
>
> Thirdly, recommend legal and administrative changes (including an information system) designed to enhance the institutional capacity to implement the proposed strategy.

The report concluded that these objectives are "extremely ambitious" and that "the fundamental question is whether it [the NUDS project] will be able to actually deliver its expected work product on time within the present scope of the project, or whether modifications [in the basic approach] will be necessary or desirable. . .". In other words, how far can an essentially "rational-deductive" kind of approach provide a sufficient basis for the development of public policy. This is not to say that scientific research and analysis are not relevant to public policy, but it is to question whether research and analysis are enough *by themselves* to provide the basis for something as complex and "political" as a national urban development strategy.

In the light of these conclusions, the Watts report made two main recommendations. First, the report observed,

> when considering the prospects for multi-sectoral development — as distinct from planning — one has always to bear in mind that the Department of Public Works is not the only agency involved, important though its achievements have been.

Thus, when it comes to project *design*, this has to be done in co-operation with other relevant government departments. As far as project *implementation* is concerned, the report makes a similar point:

> Above all, it would be desirable to modify, to the extent possible, the "top-down" approach which the whole exercise might seem to imply. . . .

Specifically, the report proposed that the relevant implementing agencies — sectoral departments and especially the various levels of local government — should participate in the work of the project. After all, these are the agencies which the project is intended to benefit; these are the eventual "customers" for its products. To work with these agencies is no more than prudent "market research" aimed at ensuring that the product delivered is what the "customer" actually needs and wants. Accordingly, the Watts report again recommended establishment of links outside the Department of Public Works, this time proposing a formal consultative mechanism with local governments through the Ministry of Home Affairs.

To summarize, the NUDS project is an ambitious attempt to co-ordinate the planning of human settlements across sectoral/departmental boundaries and to integrate that with broad national objectives. Most of the emphasis in the NUDS project has been on research and analysis as the key to determining how this should be done, but the NUDS project is also committed to making recommendations for strengthening the institutional framework for updating and implementing a national urban development strategy once it is adopted. One of the principal elements of this part of the project is the development of a computerized information system. In particular, an information system is regarded as a vital element in implementing and updating the national urban development strategy once the project has ended.

Thus, the information system was regarded more as a by-product (albeit an important one) than a component of the NUDS project. In particular, it has always been intended that the core of the information system would be provided from the data assembled by the project team. Broadly speaking, these amount to four main databases:

1. An urban facilities database at the village (*Desa*) level based on a 1980 survey (PODES/RDL) of 61,000 villages carried out by the National Bureau of Statistics (BPS); it may be possible to enhance the original survey data with data from a 1978 survey of facilities in about 2,000 urban villages (*desa kota*) that was carried out by the Department of Home Affairs (DEPDAGRI).
2. An urban services database at the subdistrict (*kechamaten*) level (INFOTRAN) based on surveys by the Department of Public Works in 1980 and 1981 of the 3,400 subdistricts and capitals of subdistricts (*Ibu Kota Kechamaten*) respectively; it is possible that these data will be enhanced with data from sectoral sources (the Departments of Education and Health, the National Electricity Corporation, etc.).
3. A multimodal transportation origin/destination database for passenger and goods (six commodity groups) movements at the district (*kabupaten*) level; the database consists of survey data from sample surveys in 1977 and 1982 carried out by the Department of Public Works (*Bina Marga*) and the Department of Communication and Transportation.
4. A miscellaneous collection of baseline data files at different levels compiled by NUDS staff from various sources and stored on about 100 microcomputer floppy disks.

In March 1984, a "Preliminary Computer System Management Plan" was prepared by NUDS project staff. The main part of this document consisted of a list of specific actions to be undertaken in order to get the computer and the information system operational. Among the most important of these actions were to be the following:

1. Make an inventory of existing project data files.
2. Prepare a preliminary design for the Information System.
3. Draft an operations plan for implementation of the proposed Information System, including management and training.
4. Create and test an initial set of data files to form the core of the Information System.
5. Add further baseline data files to the Information System.
6. Prepare a long-term training programme.
7. Develop some sophisticated software for analytical/planning applications.

Before any of these tasks could be completed, however, the plans were revised. In May 1984 (presumably in response to the comments in the Watts report), it was apparently agreed that only a "base system" version of the proposed information system would be completed under the present project (i.e., by the end of September 1985). This "base" version will provide only "for continued use of the files developed during the NUDS project". It will not

> provide the technical capabilities required to update selected original data files and to add new data files and analytic modules . . . [or] include the development of data bases and software for mapping at various scales.

An "expanded" version of the information system that will presumably do all these things was promised under a separate (but not yet approved) project.

The computer system was installed during July and August 1984. It is a Digital VAX 11/730, with 2 Mb of RAM memory. On-line storage is provided by one 121 Mb fixed disk and one 10 Mb removable disk. For off-line storage and data transfer, there is a tape drive unit (1600 bpi). There are three VT-102 video display terminals (each with printer port), one VT-240 video-graphic terminal and one LA-100 hardcopy terminal. There are three printers: one letter-quality (32 cps) and two high-speed (300 lpm) dot-matrix printers, one with graphics capability. Software consists of the VAX/VMS operating system, a FORTRAN compiler, a BASIC interpreter and a VAX database management system (Datatrieve with CDD). In addition, the system includes a Decmate II microcomputer with its own software for word processing.

In August 1984, UNCHS (Habitat) fielded a second mission (Philippe Billot) to advise the government on computer systems development in support of NUDS. The Billot report presented a review of progress towards development of a NUDS information system and concluded that significant additional inputs of personnel and equipment would be required. For as matters then stood, the report pointed out that, while "some ideas have already been developed", an actual plan for an information system "does not exist". Specifically, the report had this to say:

> *System Database*: the report drew attention to weaknesses in the data files intended for use in the information system:
>
> - documentation: many of the files lack adequate information on sources, definitions, etc.;
> - accuracy: some of the survey questionnaires were enumerated by officials while others were just mailed out;
> - consistency: e.g., not all of the files used the same criteria for urban/rural classification;

- completeness: some of the surveys were not completed;
- age: some of the data were then already five to seven years old; and
- frequency: e.g., some of the data files are unlikely to be updated until 1990 or later.

System Hardware: the report pointed out that the computer can support up to eight terminals in its present (2 Mb) configuration, but that there is no provision for connection of remote terminals or microcomputers.

System Software: the report drew attention to the lack of software suitable for creating an information system and to potential difficulties involved in transferring data files to and from the database management system (Datatrieve) running on the VAX.

The Billot report made a number of recommendations for the future:

1. Establish a Data Management Group under the direction of a Data Administrator supported by a Data Processing Manager with a computer-science background.
2. Establish a uniform spatial referencing system for integrating all the various data files.
3. Acquire a general-purpose statistical software package.
4. Examine the prospects for integrating the Information System with project management (specifically the INTAL system) and/or mapping and graphics.
5. Prepare a comprehensive training programme that would cater to the needs of both serious and occasional users.
6. Be prepared to buy additional computer capacity (2 more Mb of RAM), software (for project management), peripheral devices (digitizers and plotters) and consultants (for training, applications programming, systems development, etc.).
7. Do not let existence of the minicomputer system retard the use of microcomputers; concentrate on making it easy for both kinds of computers to enhance their respective capabilities.

Total additional cost for these improvements was estimated by the report at between $200,000 and $250,000 — nearly twice the cost of the original system.

In the months since the Billot report, installation of the computer system has been completed, and it is now in full operation. According to the computer staff, there are some six to eight regular users relying mainly on FORTRAN and (to a lesser extent) Datatrieve. There are about 40 Mb of user files on tape, although at least half of these are probably secondary data files (i.e., subsets or smoothed versions of other files) or automatic file backups made by the system. In particular, none of the four main databases proposed for the information system (see above) has yet been transferred to the VAX, nor is there any software ready on the VAX capable of turning the databases into a working information system.

To summarize, the Billot report had much the same to say about the proposed information system as the Watts report had said earlier about the project as a whole, i.e., that its aims are rather "ambitious" and that "the fundamental question is whether it will be able to actually deliver on time . . . or whether modifications will be necessary. . .". Moreover, Billot's reasons for making this assessment are essentially the same as those advanced by Watts. First, there has been a tendency to think of the information system in the abstract, with little thought of who will eventually use it and for what purpose. Secondly, there has been a tendency to think that the information system can be implemented "from the top down", with little reference to the specific or local contexts where it will ultimately be used and updated.

At this point, therefore, the prospects for a full-scale national information system on urban development arising out of the NUDS project have to be rated as slim. The project team effectively reached the same conclusion when it modified the specifications for the information system. Two proposals were made: one was to confine the system database to data already available in project data files, and the other was to omit any capacity for updating or expanding the database. The first of these suggestions is sensible enough in the circumstances, but the second would seriously affect the viability of the information system. If an information system cannot be added to or updated, it can hardly serve as a means of monitoring performance. Data alone do not make an information system: one of the most important features of an information system is its maintenance, i.e., the process whereby it is kept up to date. If this capacity is omitted, the information system is effectively "dead" before it ever goes into operation.

Thus, rather than preserve the *scope* and compromise the *nature* of the information system, it would be better to do the opposite. Rather than try to build into the proposed information system *all of the available data* at the cost of reducing the capabilities of the system, priority should go to selecting the best and most relevant data and using it as the basis for creating a viable *information system* that really serves and is served by planners in the Department.

This means that the source of data for the information system would still be the project databases, but selection from that source would be based on (a) assessment of users and their information needs and (b) the likelihood of the data being kept up to date as time goes by. In this way, the needs and capacities of users become as important a part of the information system as do the data themselves.

For the project, this implies broadening the scope of work involved in setting up the information system. No longer is it enough to ensure the databases are available; now, it is also necessary to assess user needs and data resources. At the least, therefore, design of the information system will entail the following steps:

- Identify and consult with intended users of the information system with a view to identifying potential uses, including data requirements, computer processing and computer access entailed in these uses.
- Identify at the earliest possible time those elements of the project databases most appropriate for an information system and concentrate project resources on refining them into a core database for the information system.
- At the same time, identify sources of data and processes of data collection to be used for updating and expanding the core database and design appropriate management mechanisms for ensuring that this process occurs on a regular and frequent basis.

As far as the computer system itself is concerned, the important thing is to make the VAX as accessible as possible to potential users, including the growing network of microcomputers within the Department and in the government as a whole. Information systems, more than many computer applications, depend on communication, both to get data in and to get results out. As a result, there may be a tendency to feel that the existence of the VAX ought to set some sort of standard within the Department. This may give rise to demands for "VAX-compatibility" whenever anyone wants to buy a computer; or there may even be an outright prohibition on other computers as long as there is spare capacity on the VAX. This would be a mistake. Computers are not "all the same underneath"; different computers serve different needs. People cannot be coerced into using the VAX if they judge it inappropriate for their needs and capacities. Indeed, people are more likely to be "recruited" to the VAX and to use of the information system through preliminary experience with the friendly environment of microcomputers than if their only choice is to go directly to the VAX.

In particular, steps should be taken to make the VAX capable of communicating easily with microcomputers, so that as much of the on-line data-processing as possible can be carried out locally in the friendly and flexible environment which microcomputers provide. Meanwhile, the VAX can be reserved for computationally intensive uses. The objective should be to make the VAX compatible with microcomputers, rather than the other way around. Thus, the VAX will need programming for three main purposes:

1. to create the programme (presumably under Datatrieve) to operate the information system;
2. to provide user-friendly, menu-driven "shells" for the above programme so that it can be used by non-specialists; and
3. to make the VAX operations "transparent" to the microcomputer user (for example, if file conversion of some kind is required in order to make VAX files readable with a standard microcomputer programme, the VAX should be able to provide a simple menu-driven programme to do the conversion).

Finally, there will be a need for training. Its nature will depend, of course, on the eventual design and expected use of the information system. As a general rule, however, it is important to recognize the need to serve both casual and specialized users. It may be appropriate to send the latter on "VAX courses"; but it would be a mistake to ignore the needs of those who may be unwilling or unable to undertake that kind of training. If the information system is meant to serve users throughout the Department, those users have to be among the clients for a training programme, and, if access to the system can be provided from microcomputers, that is probably where training for the casual user should be carried out rather than on the VAX itself.

Case 10

Top-down versus bottom-up design (India)

Improving the information base for human settlements planning in India has been an objective of government policy for a number of years. Now, with advances in the field of low-cost microcomputer technology, the prospect of realizing some form of national information system for urban and regional planning has become realistic. Nevertheless, the task of co-ordinating and automating information for planning in the nearly 3,500 municipal jurisdictions (cities, towns and districts) in India remains a daunting one — to say nothing of doing so in the more than 576,000 villages that make up the rural areas of India.

The responsibility for establishing an information system for human settlements planning has been assigned to the Town and Country Planning Organization (TCPO) of the Ministry of Works and Housing. After the idea of such a system was first proposed at the annual meeting of the Chief Town Planners of the States and Territories in 1976, TCPO contacted the United Nations Centre for Housing, Building and Planning (the forerunner of UNCHS Habitat) to ask for advice on the feasibility of establishing an information system at various levels of government to assist in urban and regional planning in India. The report of the mission (which visited India in January 1977) concluded that such an information system was indeed feasible in India but drew attention to a number of potential constraints. The basic conclusions were (UNHCBP 1977):

- No operational information system can truly meet all the needs of all possible users; so it would be important to set priorities, identify "vital data elements" and contemplate different (albeit integrated) systems for different levels of government.
- The quantity of planning data available in India generally was thought to be adequate to support information systems, but its quality was judged to be uneven; in particular, its timeliness was assessed as only "fair to poor".

- Attention was drawn to the crucial role of electronic data-processing in creating information systems; the report emphasised the need for facilities (including hardware and software) and training.
- Finally, the report recommended a demonstration project in an urban area (such as Delhi or Madras) in order to test various designs and provide experience to TCPO staff as well as planners and other users.

The following year (1978), TCPO prepared a formal "Scheme for Establishing Information System for Regional and Urban Development" and submitted this to the annual meeting of the Chief Town Planners. Out of this meeting came a recommendation that "statistical cells" should be created in the Town and Country Planning departments (or their equivalents) in all states and territories as a prelude to establishing a national information system.

In August 1979, TCPO convened a two-day meeting in Delhi of 45 experts on information systems drawn from across India. The aim of the meeting was to review proposals for what had by then become known as the Urban and Regional Information System (URIS) and to discuss how (in the words of the Secretary of the Ministry of Works and Housing) "to make a start on a modest scale and to progressively improve the system". The meeting recommended that the Ministry should proceed with establishment of URIS in association with the state and territorial governments and that a Steering Group of experts should be appointed to provide continuing advice on development of the System. The Steering Group was duly appointed and began work in November 1979.

In its first report on information systems for urban areas (TCPO, April 1981), the Steering Group laid the foundation for a "top-down" approach. The report set out what it called a standard "framework and data format" consisting of 22 main "variables" and about 274 "sub-variables" reflecting three principal "components" of the "urban subsystem": the social and economic environment, the physical environment, and services and amenities. Beyond this, the Steering Group offered no specific criteria for its selection of data items, saying only that it had been guided by the need to include "aspects considered crucial" to planning at the local level and the need to reflect "broad relationships between various components" to assist decision-makers at high levels.

In its second report (TCPO, March 1983), the Steering Group dealt with information systems at the regional rather than urban scale. The report proposed a data format for what it called a "generalised Regional Information System" consisting of nearly 400 "sub-variables"; but this time the Steering Group made no special claim for its list of variables, noting merely that it had listed "all the information gathered by TCPO" for planning purposes and "critically examined" it before coming up with its proposed format. Moreover, the second report also acknowledged the difficulty of incorporating both area-specific and location-specific data in an information system and underlined the need for a consistent approach to geocoding. Accordingly, the Steering Group recommended that a number of different information-system designs should be created and tested in "selected regions of the country [and] at different scales of regions with a view to evaluate [*sic*] the merits and limitations" of each.

With the encouragement of the Steering Group, TCPO began lobbying state and territorial governments with a view to securing agreement to collaborate on a number of pilot studies. In June 1981, the government of Tamil Nadu agreed to run a pilot study in Chengalpattu, a small town of just under 50,000 people located about 50 km from Madras. The next year, a second pilot study was launched, this time in collaboration with the State of Gujarat and focused on the town of Anand, with a population of about 84,000 located about 100 km from Ahmedabad. However, the purpose of the two pilot studies has been confined to "validating" the 274-item "data format for urban systems" proposed by the Steering Group. There has been no attempt in either case to examine other formats or actually to implement and evaluate a working information system.

Although neither pilot study is yet complete, their apparent conclusions are not surprising: while useful comments can be made about the suitability and availability of certain kinds of urban data, it has proven difficult to "validate" the notion that one particular "data format" is in some sense ideal.

For one thing, as the Steering Group itself acknowledged, some important factors had to be left out of its proposed "data format". For example, according to the first report (TCPO, April 1981), "very little information is yet available about air and water pollution"; all such data were therefore omitted. Thus, the report continued, the proposed "format may not suffice where more vigorous analysis of relationships is required [as], say, in the context of formulating a city plan". Data on industrial and commercial activities, for example, are limited to physical data, i.e., data on the number of people employed, the number of industrial estates and the types of land use in an urban area. There are no corresponding financial data, such as data on payroll, investment, turnover, value added and taxes paid. For some planners, data like these are indispensable.

At the same time, there are also serious problems of ambiguity and vagueness in the "data format" proposed by the Steering Group — problems which make it seem doubtful that the "data format" was intended to be definitive rather than simply illustrative. In the Steering Group's first report (TCPO, April 1981), for example:

(a) The terms "variable" and "sub-variable" refer to different data sets. Sometimes the terms refer to single data values; at other times they refer to vectors or matrices. Sometimes it is not clear what is intended.
(b) The "data format" does not always indicate what kind of data are intended and what units of measurement are implied. For example, sub-variables such as "rainfall", "temperature", "humidity", "wind velocity", "physical features", "slope", "topsoil" and "water" (surface and subsurface) may imply quantitative or qualitative data, absolutes or ratios, stocks or flows, mean values or ranges, etc.
(c) There is little explicit discussion of how far (if at all) historical data are to be retained in the system. The report pays some attention to the question of updating data (suggesting a suitable "periodicity" for each sub-variable); but it says nothing about whether new data are to be added to or are to replace old data.

Of course, it can be argued that these questions will all be resolved when the system is implemented, but that is precisely what makes it so difficult to "validate" a "data format" in the abstract. The fundamental question this poses is how far (if at all) it is reasonable to try to design an information system without reference to the context in which it is going to be used, i.e., who will use it, how it will be used and what it will be used for. It is hard to see how any data format can be validated without reference to questions such as these.

In the end, opinion on the Steering Group's proposed "data format" is likely to remain divided. According to some, the format is not extensive enough. For example, the data may be sufficient for policy-making at national or state level but not for planning at the local level. Thus, the second report of the Gujarat study (TPVD, 1985) concluded that "at best the list of variables can help one to be informed about a town and its characteristics but [they] hardly meet the needs of the [town] planner". According to others, the proposed data format is already too extensive, especially for this stage of its development. According to another recent report, there was a strong recommendation from a national workshop in 1983 that a less ambitious design be adopted in order to get an information system in operation as soon as possible:

> In view of the urgency of the need to have an Urban Information System . . . so that it can provide meaningful inputs in the preparation of the seventh and subsequent Five Year Plans . . . the information base of the national sub-system may not be very

> comprehensive either in terms of data input or the proposed output structure. It can start off by targeting . . . a few well defined output parameters (indicators) that are generally used in the formation and evolution of urban plans, including plans currently undertaken by the concerned Ministries in the centre and the states.
>
> (ICOR, 1985)

Similarly, in its third report, the Tamil Nadu study in Chengalpattu called for emphasis on analysing the constraints of computer processing on the design of information systems and on identifying likely wants and needs of users (DTCP, 1984).

In the meantime, TCPO has also been developing a number of modest "prototype" information systems for use at national and state/ territory levels. One such system, URDIS, is already operational and several others are at various stages of planning and implementation.

> URDIS (Urban and Regional Document Information System) is an inventory of approximately 1,200 urban research studies by university and governmental bodies in India since 1975. For each study, URDIS contains information on author, subject, source of funding, stage of completion and nature of output; there is no summary or abstract of findings. The inventory is limited to formal research studies and (so far) does not include studies or projects carried out as part of the planning process. URDIS has been in operation for about a year and a half, during which time it has doubled in size. URDIS runs on the mainframe computer at the National Informatics Centre (NIC). Data entry is done through a terminal at NIC, and printouts are provided on demand (which so far has meant about once every six months). There are no facilities for interactive use. Plans for expansion include publication of an annual "directory" of urban and regional research.
>
> A second national information system under way in TCPO is MONIS. MONIS (Monitoring Information System) is designed to help monitor implementation of the national programme for Integrated Development of Small and Medium-Sized Towns (IDSMT) for upgrading infrastructure such as roads, drains, sewers, markets and clinics. MONIS will contain background data on each of the towns in the programme as well as data on financial disbursements and performance relating to each separate scheme. When it is fully operational, MONIS is expected to cover about 1,000 different schemes in 236 different towns. MONIS, like URDIS, will run in batch mode on the main NIC computer in Delhi.
>
> A third information system in TCPO is planned to support a national programme on Environmental Improvement of Urban

Slums (EIUS). This is a programme aimed at improving physical conditions in urban slums through financial support for schemes to provide potable water, community baths, pavements, drains, sewer lines, street lighting, etc. The information system is intended to record slum conditions and programme performance by State/Territory throughout the country.

A fourth information system planned by TCPO is an Organization and Manpower Information System (OMIS). OMIS will be designed to provide an inventory of manpower and resources in planning agencies at all levels of government. This information will be useful to TCPO in determining the nature and level of training needs and will provide the raw material for an annual "roster" of planners in India.

A fifth information system planned by TCPO is a Land-Use Information System (LUIS). Land-use data for urban areas is normally collected in a systematic way only in response to the needs of preparing a masterplan. Only about 500 of the 3,500 municipal jurisdictions have masterplans, and some of these are well over 10 years old. Nonetheless, TCPO would like to collect all the available data into a single information system.

Although only the first of these systems is currently operational, most of the others easily could be, as soon as suitable hardware and software become available. For the time being, the only facilities available to TCPO are those of the National Informatics Centre (NIC): a Cyber CDC 170/730. Thus, processing is remote, customized and batch — which is not the most appropriate environment for information systems. On the other hand, with an interactive and user-friendly computer system at its disposal, TCPO could soon have half a dozen prototype information systems in operation.

In summary, there has been considerable activity by TCPO towards development of a national urban and regional information system since the original proposal and the UN-sponsored review nearly a decade ago. Two different kinds of feasibility studies have been carried out — two pilot studies in collaboration with state governments in Chengalpattu and Anand and several "prototype" systems within TCPO. The question is how far all this represents real progress towards an information system.

To most people, an information system means a tool for managing data and presenting it in a form suitable for decision-making. As such, an information system has three main components: (a) inputs (raw data, usually numbers but occasionally text); (b) processing (equipment and processes for storing and manipulating inputs and converting them into desired outputs); and (c) outputs (display of the information desired in the form of tables, statistics, projections, models, maps, charts, diagrams, etc.).

The second characteristic of an information system is the level of decision-making at which it is used. In some cases, an information system may be meant for just one basic use, such as monitoring the performance of a particular project or programme. In other cases, an information system may be expected to serve many uses, including management, planning, analysis, evaluation and policy formulation. Naturally enough, the more varied the uses of an information system, the more difficult it is to design — to the point where an information system with no specific uses (and thus meant to serve all possible uses) is very difficult to design indeed.

The third feature of an information system is the scope of the information it produces. Here again, information systems can be designed to cater to the specific (e.g., purely financial information) or to the general (e.g., a broad range of socio-economic and environmental information). Again, too, the broader the scope of the system the more complicated it is likely to become, although the scope of the data fed into a system effectively puts a limit on the scope of the information it can produce.

Now, consider how these three dimensions apply to the proposed national urban and regional information system. First, the system is going to involve three distinct activities: gathering data in various forms and from various sources (input), processing data (manipulating the data in a computer or computers using appropriate software) and producing information (output) for planners in the form or forms they need. These three stages in the functioning of an information system are illustrated in the horizontal dimension of the diagram in Figure 3. Data inputs are represented by the arrows on the left, data-processing by the large rectangle in the middle and information output by the thicker arrows on the right. Secondly, the proposed information system is intended to be used at all three levels of government, i.e., Union, state/territory and local levels. This is represented by the vertical dimension of the diagram in Figure 3, in particular by the three labelled boxes within the large rectangle. The arrows between these boxes represent "vertical" data flows between the different levels of the information system. Thirdly, the scope of the proposed information system is intended to be quite large (at least eventually). This is difficult to illustrate in a diagram that is only two-dimensional; but think of the "width" of the box in the middle as representing the breadth or scope of the information system.

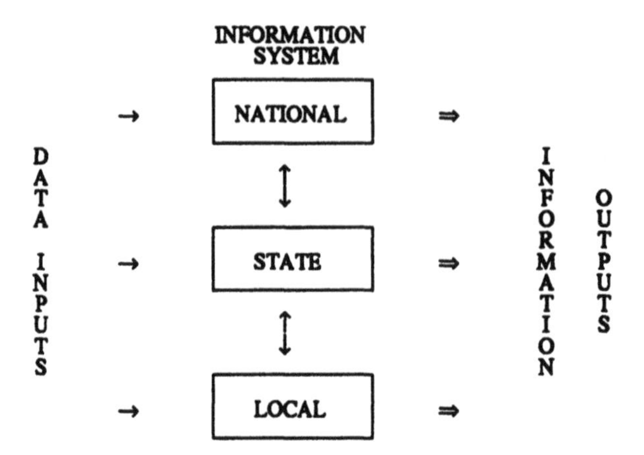

Figure 3 Schematic diagram of a multi-stage, multi-level, multi-use information system

By way of further illustration, consider this. Union planners may want an information system to help formulate national policies and monitor national programmes such as the programme for Integrated Development of Small and Medium-Sized Towns (IDSMT) or Environmental Improvement of Urban Slums (EIUS). Similarly, state planners may want an information system they can use for preparing area (structure) plans for towns and districts and for monitoring state programmes (such as sites-and-services projects or neighbourhood improvement projects). For their part, local planners in metropolitan, municipal and district agencies may want an information system they can use for detailed area planning and project management at specific sites.

Even these examples represent only a subset of all the possible uses of an urban and regional information system. Thus, information systems can soon become quite complex. Data input, data-processing and information output are all closely related to one another — a relationship enshrined in the well-known dictum, "Garbage In, Garbage Out". Similarly, the needs of different users are often related; yet they are also different. They are related in the sense that the same basic data about population, housing, employment, infrastructure, services, etc. are important for planning at all three levels of government; similarly, local project management can provide data for programme evaluation at state levels and programme data from the state level is valuable for national programme assessment; but their respective information needs are also different. Senior levels of government may want more aggregated data than local planners; local planners, on the other hand, may want more detailed information about a smaller number of variables than do central planners.

The implication of all this is that an information system may in reality be a "system of systems" rather than a single, all-inclusive system — that is, an integrated, overlapping network of separate systems. These overlapping systems could share some of the same data structures, some of the same procedures for manipulating the data and some of the same output capabilities; but they could also have their respective peculiarities reflecting the management and planning needs of their particular users.

The "common core" of data, software and output will provide the system with the consistency and efficiency that justify its establishment. In particular, the "common core" will help ensure that data can be passed back and forth between different parts of the system, thus reducing duplication and waste. At the same time, the variety of subsystems within the information system as a whole will provide the ability to respond to different kinds of user needs and to adapt easily to changing needs over time.

The essential feature of information systems (indeed, of all systems) is that their parts are dynamically related to one another. For example, the kind of data that goes into an information system — its reliability, scale, periodicity, timeliness, etc. — affects what kind of data-processing can be done and (thus) what kind of information can be produced by the system.

Similarly, data-processing affects what kind of data can usefully be put into the system and what kind of information can be got out. Of course, procedures can in principle be devised for manipulating any kind of data; but in practice there may be critical constraints on resources, timing, user skills, etc. that severely limit what can actually be done. This is particularly true of computerized information systems that are intended for direct ("interactive") use by large numbers of essentially unskilled users. Here the possibility of building a system around much more than a few well-established, "user-friendly" packages of a generic nature is doubtful.

Finally, in an information system, the nature and form of the desired output affects both the data inputs and the procedures required for manipulating them. If the information system is intended to produce maps, obviously the data must include a spatial reference code and the procedures must include an ability to generate maps. Again, if the system is intended to perform economic modelling, appropriate data and procedures for that kind of output must be provided.

This interrelationship between data and their intended use is what distinguishes information systems from databases (such as a census, an archive or a library), where data are gathered largely for their own sake, independent of any specific uses. In an information system, particularly in a computerized information system, data, software and use are all interdependent.

The most important lesson of the TCPO experience, therefore, seems to have been to underline the fundamentally dynamic and interrelated nature of information systems. For this reason, their design involves co-ordinated choices about data collection, data-processing and information output. Thus, design cannot just follow a linear process — first, identify the data requirements; then, when that is done, collect the data and put them into a computer; then, finally, print out the information required — as though each step were independent of the other two. Information systems have to be designed iteratively — with questions of what data, how they are processed and what results are desired all being addressed simultaneously — so that incremental decisions about one can be constantly co-ordinated with similar decisions about the others.

The goal of a national urban and regional information system is fundamentally a very ambitious one. What it implies is nothing less than a single, integrated source for all (or at least most of) the information considered necessary for urban and regional planning at Union, state/territory and local levels. It is still an open question whether the goal of a comprehensive national information system can in fact be realized. It can be argued that no such system is operational anywhere in the world. In any case, two other conclusions would seem warranted. First, creating an information system on the scale proposed is clearly going to be a long-term undertaking. Secondly, in the circumstances, it would be prudent to have some less ambitious goal or goals in mind as a fall-back in the event that the comprehensive one proves elusive.

Thus, as a general principle, the most sensible approach to designing large-scale systems is through establishment of a number of small subsystems. These subsystems should be consistent with the requirements of the overall system, but the subsystems should also be capable of working independent of the overall system and should have a value and utility of their own. Thus, while or, even, if the hoped-for synthesis of the subsystems into an overall system remains unrealized, the investment already made is not wasted. At the very least, some users of the subsystems will have better information-management capabilities than they did previously. Moreover, there is another purely practical reason for insisting on a visible pay-off from information systems: people who benefit from an information system have a motive for keeping it up to date. The best way to ensure the success of an information system is to design it so that its output is useful to those whose co-operation is necessary for keeping it up to date. The more effort that has to go into compelling people to provide data for or persuading people to make use of an information system, the less worthwhile the whole exercise begins to be.

Thus, the main conclusions can be summarized as follows:

1. A national urban and regional information system should be thought of not as a single system but as a "system of systems", each designed so that the usefulness of its output is sufficient to motivate those responsible for keeping it up to date.
2. The various feasibility studies carried out by TCPO appear to show that data availability is not a critical problem; that is, there are problems with data availability but they are not serious enough to threaten the fundamental viability of some kind of urban and regional information system.
3. The critical factors are going to be access to appropriate computer facilities, including software, on a widespread basis and provision of training necessary to enable planners to take advantage of it.
4. The next step should be design and implementation of a multi-stage, multi-level demonstration project focused on a specific use, including supply of equipment and provision of training.

In general, these conclusions reinforce many of those reached in the 1977 review by the United Nations Centre for Housing, Building and Planning. However, in the intervening years, development of the microcomputer has brought about a revolution in data-processing, thanks to a whole new generation of low-cost equipment and user-friendly software. This has put information systems within the reach of almost anyone. Consequently, the emphasis in information-system design has shifted from trying to include "every datum required for any purpose whatsoever" — as recommended recently (Davies 1984) to the Madras Metropolitan Development Authority — to being selective about data-gathering and to providing flexibility in the way data can be used.

These changes are reflected here in the emphasis on building up a national information system out of small, functioning subsystems. In order to do this, interactive access to computer facilities (hardware and software) has to be provided to those who use the system and those on whom the system relies for updating. At the same time, the wide accessibility of computer facilities (albeit simple, low-cost facilities) coupled with the user-friendly nature of the software will combine to make training fast and easy; indeed, much of it can be done on the job.

Case 11

Data banks and information systems (Philippines)

The Philippines Ministry of Human Settlements was created in June 1978 by Presidential Directive No. 1396. According to the Directive,

> Human settlements is a prime concern and a major approach of our government in our people's quest for a better quality of life. Human settlements will always be the primary habitat of the human being, the intimate milieu of his civic identity and the immediate environment of his creative experience.
>
> (*1983 Annual Report*, p. 45)

The Ministry consists of three main Groups (a Shelter Development Group, a Livelihood Development Group and a Community Development and Services Group), three Support Groups (a Strategic Planning Group, a Regional Operations Group and a Communications Department), and some 32 "attached agencies", including the Human Settlements Development Corporation (HSDC), Human Settlements Regulatory Commission (HSRC), National Housing Authority (NHA), National Housing Corporation (NHC), National Home Mortgage Finance Corporation (NHMFC), Home Financing Corporation (HFC), Home Development Mutual Fund (which administers the Pag-IBIG scheme) and Bliss Development Corporation.

Human settlements planning in the Philippines is fundamentally the responsibility of local governments. A World Bank report summarizes the organization of the urban sector in the following terms:

> The government structure . . . is centralized with national government selectively delegating power to lower levels — provinces, cities, municipalities and *barangay* (neighborhoods). . . . Provinces historically have not been strong political or administrative entities and most local administration has come from the 1,484 municipalities into which the provinces are fully subdivided. All areas . . . are incorporated and municipalities contain both urban and rural areas. There are, in addition, 60

chartered cities which are more highly urbanized than municipalities though also contain rural areas. . . . Cities and municipalities are headed by Mayors who are elected officials (except in Metro Manila) and are responsible to locally elected councils. . . .

All local governments have authority to set and collect charges and taxes, prepare budgets, hire staff and invest in and manage local services and enterprises. . . . In addition to the construction, operation and maintenance of local roads, bridges, drainage and solid-waste management [facilities], local governments are responsible for the provision of education and health facilities for which they receive some central-government finance. Economic enterprises run by local government include public markets and slaughterhouses with fees and rents being charged to lessees. Water supply is administered by autonomous local Water Districts which charge user fees. . . .

(*Staff Appraisal Report*, 1984, pp. 1-2)

The principal power and duty of the Ministry of Human Settlements in respect of local governments are set out in a Presidential Letter of Instruction (LOI No. 729):

> The Ministry of Human Settlements shall prepare or cause to be prepared land use plans and zoning implementation and enforcement guidelines for urban and urbanizable areas. . . .
>
> Local governments are required to submit their existing land use plans, zoning ordinances, enforcement systems and procedures to the Ministry for review and ratification.
>
> All other communities without land use plans and zoning implementing guidelines shall prepare said guidelines, and may call upon the Ministry of Human Settlements for planning assistance, etc.

The specific duty of "review and ratification" of town plans is carried out by the Human Settlements Regulatory Commission on behalf of the Ministry, while the Ministry itself concentrates on providing planning assistance to municipalities.

One of the goals of the Ministry of Human Settlements has been to maintain a national information system for human settlements planning. In fact, the idea of a Planners' Data Bank goes back to the mid-1970s. According to the Executive Director of the National Census and Statistics Office,

> The plan [for a Planners' Data Bank] calls for harnessing available information systems and computer resources and interlinking them

> to provide an integrated data base for national policy-makers and planners. Programme-level data banks are . . . to be established in . . . government agencies [involved in] national development . . . and in strategic regional centers in support of a national-level data bank. An essential feature of this plan is the use of prescribed standards in classifications, processes, products, services, and procedures [in order to promote] comparability and interchangeability of information. . . .
>
> (*User's Manual*, 1978, p. i)

In its original form, the Planners' Data Bank was to have three main components which would "eventually be linked to one another so that maximum utilization can be made of the data":

- a numerical database consisting of data on demography, physical features, productive sectors, social characteristics and infrastructure;
- a geographical database consisting of data on climate, soil, geology, slope, drainage, water resources, land use, vegetation, social services and infrastructure; and
- a project database consisting of data on types and locations of development projects, as well as their proponents, sources of funding, implementation schedules and cash requirements.

In summary, the aim was to create a single, comprehensive database on human settlements capable of meeting the needs of policy-makers, planners and managers at all levels of government.

The Data Bank consists principally of three databases which (according to a draft report currently in preparation) together provide "a continuing data series on the availability of services, facilities and infrastructure" at the municipal level to meet 11 basic needs — "food; shelter; clothing and cottage industries; water; power; livelihood (economic base); education and culture; health, nutrition and welfare; sports and recreation; mobility and ecological balance" (MHS 1985, p. 2). The three databases are derived from questionnaire returns from local-government authorities in each of the approximately 1,500 cities, towns and municipalities in the Philippines in 1978 and 1982, supplemented with data on population, housing and employment from the 1970, 1975 and 1980 censuses. Each database is approximately 10 to 15 megabytes in size.

The Planners' Data Bank has provided the basis for four main outputs to date:

Settlement profiles

The first two editions appeared in 1978 and were nominally based on 1975 data; however, "a majority of the information refer[red] to earlier years". Similarly, the third edition, which appeared in 1980, was nominally based on 1978 data, but contained some data from earlier years. The fourth edition is expected in early 1986; it is nominally based on 1982 data, but will contain some data from the 1980 national census.

Planning atlases

Planning atlases are being produced in three series for every region, province and municipality. So far, atlases have been completed for all regions, most provinces and about three-quarters of the municipalities. More than half of the maps at the regional level are computer-generated, but few at the provincial and none at the municipal levels are computerized.

Town planning

Approximately 150 municipalities are required to prepare new town plans each year. For each of these, the Planners' Data Bank is used to generate a set of 15 reports on current and projected growth in population, employment, housing, etc. The reports provide the data and take the form recommended in the Sèctoral Guidelines for town planning prepared by the Ministry.

Services delivery profile

To provide a measure of the development of human settlements throughout the country, a set of 11 "indicators" (one for each of the Ministry's basic-needs sectors) was constructed for each municipality and aggregated to provincial/regional levels. This profile has just been completed using the 1978 database; use of the 1982 database awaits completion of the corresponding Settlement Profiles.

In addition, there are three other database projects currently in progress or under consideration, the effect of which will be to extend the idea of a Planners' Data Bank down to the *barangay* level (of which there are some 42,000 throughout the country). These are:

Barangay Profiles Project

A questionnaire was recently sent to Ministry officials in local Plan Implementation Teams for administration to a sample of 1,500 *barangay*. The questionnaire is designed to elicit basic data on physical, socio-economic and political conditions in each *barangay*. Presumably, the objective is to be able to create *barangay* "profiles" analogous to the Settlement Profiles at the municipal level.

Community Development Indicators Project

In 1984, the Ministry of Local Government conducted a survey of all 42,000 *barangay* and obtained a large database (estimated at 500 megabytes) of some 440 variables for each *barangay*. The Ministry is undertaking a pilot project focused on 54 of the variables and 1,350 of the *barangay* (those in the province of Pangasinan) to develop a set of community development indicators analogous (presumably) to those developed for the Service Delivery Profiles.

Barangay Development Classification Project

In 1984, the Community Organization and Development Group of the Ministry sent a questionnaire to all 42,000 *barangay* asking for simple answers (yes/no) to various questions about local services and facilities. The result was a small database (about 5.6 Mb) of 40 variables for each *barangay*. Current plans are to process the data manually and to decide subsequently whether to "bank" the data in the computer.

The original 1975 database for the Planners' Data Bank was created on a UNIVAC 1100. Now it runs on a DEC PDP-11/34 (with 512 Kb RAM, a 10 Mb disk cartridge and 80 Mb diskpack, a 45 ips tape system, nine assorted terminals, a Tektronix 4010 graphics display, and a Centronics 6300 lineprinter). The Planners' Data Bank has been in operation in the Ministry of Human Settlements for about seven years. As such, it represents an early and significant commitment to the role of computerized data management in human settlements planning and management. Only a few countries have as much experience as this to draw on.

When the Planners' Data Bank was originally created, there were two key assumptions made. First, it was assumed that there is a fixed and finite quantum of data that is necessary and sufficient for planning at all levels of government. Secondly, it was assumed that information is an input to the planning process but not really part of the process itself. Thus, the function of the Data Bank was conceived almost entirely in terms of its being a purveyor of standardized tables and maps of "key" planning data. There was no attempt to identify specific users of or clients for the Data Bank, since the intention was that the Data Bank should serve everyone. Similarly, in the design of the Data Bank, there was no attempt to determine what data users would find useful, since the importance of data was regarded as a matter of principle, not as a matter of empirical convenience.

In the years since the Data Bank was originally created, planners have begun to question these assumptions about the role of information in planning. For one thing, no one has yet succeeded in creating a truly comprehensive planning data bank. Secondly, there is a growing feeling that information is more than just something for planners to insert in their documents at appropriate places. Instead, information constitutes the environment in which planning actually occurs — the context in which ends and means are created, evaluated and selected. Information, in other words, is not such a one-dimensional commodity as was once assumed.

What this means is that the goal of a single, integrated data bank on all facets of human settlements planning may be misplaced. Indeed, it may even be counterproductive, if the search for the "perfect" data bank is allowed to divert attention and resources from more mundane and painstaking achievements. For the fact that a single, all-purpose data bank is unrealistic does not mean that nothing can be done to rationalize the flow of information for planning. A lot of data can indeed be usefully shared among planners in different sectors or at different levels of government. Moreover, this may well result in significant cost savings and/or improvements in productivity. However, the critical issue is precisely this: that planning databases are not idealized "good things" that can be designed and implemented on abstract principles. Planning databases have to be built "from the ground up" and justified on the basis of what specific users need and what are the best available data to meet those needs.

A second fundamental implication of this more sophisticated view of the role of information in planning is that planners need both to "have" data and to be able to "interact" with them. In other words, providing data to planners is only half the battle; if data are to be useful for planning, planners need to be able to analyse and manipulate the data. The advantage of a computerized database lies not just in its ability to contain large quantities of data. What is just as important is the fact that the computer can rapidly retrieve, process and present the data not just in one or two standard formats but in as many different ways as the planner can think of. Thinking about data in different ways lies at the heart of the planning process. So a planner's information system needs not just to contain data but also to enable him or her to manipulate them effectively.

In some respects, this distinction between information and the capacity to use it for planning corresponds to the distinction between plan-making and planning. Both have to do with the difference between planning as a "process" and the plan as a "product" or "output". Faced with the complexity and dynamism of a modern society, data and plans are merely means to an end. Effective planning lies in the ability not just to create the means but also to adjust them to changing circumstances. In the words of the Ministry's own *Guidelines* (1982),

> Town planning is a dynamic process. It entails continuing efforts on the part of all those concerned and does not end with an approved town plan; hence, this manual by necessity reflects flexibility and remains open to further refinement and modification.

The following paragraphs give some detailed comments on the flexibility and adaptability of the Planners' Data Bank:

Scope and coverage

Since the Data Bank was not designed to meet any specific set of needs, it is difficult to know what criteria to use in assessing the scope and coverage of the Data Bank. It is easy to think of data that could have been included but were left out; for example, there are no economic or financial data in the Data Bank. Similarly, it is not hard to question some of the data that were included; for example, what particular (planning) purpose is served by identifying the municipal location of radio stations? However, it is difficult to substantiate these criticisms without being able to relate them to some specific uses of the Data Bank. If intended users were not identified at the time the Data Bank was established, an alternative would be to do some "market research" now and find out who is using or who might be able to use the Data Bank and then to see what improvements could be made to enhance those uses. As matters now stand, the Data Bank is essentially a tool in search of a use, a solution looking for a problem. Naturally, this makes it difficult to assess how good a tool or how effective a solution it is.

Sources, timeliness and updating

The Data Bank is based on two sources of information — the national census and a periodic questionnaire to municipalities. There are no built-in procedures for continuous updating, even on a partial basis, by either the National Census and Statistical Office or the municipalities. In both cases, a specific decision is required to proceed with an update — decisions which, as in the case of the 1985 "mid-term" census, are by no means automatic. Thus, "current" data in the Data Bank are already three to five years old, and the tabular outputs (the Settlement Profiles) are not yet available. This situation is not satisfactory, particularly from the point of view of town planning. There comes a point where the value of information for planning purposes no longer exceeds the cost of collecting and processing it. Municipalities may wonder about the value of providing data to the Ministry only to find that essentially the same data are given back to them three years later ostensibly to help them with their planning — by which time they may well have collected more up-to-date information of their own.

Accuracy and aggregation of data

Returns from the municipal questionnaires have been very high (reportedly at the 90 per cent level in 1982), but there is little attempt to verify the returns by sample checking in the field or internal error-checking by the computer.

As a rule of thumb, planners need data disaggregated to at least one level below the one they are planning for. Thus, national planners need at least regional data, regional planners need at least provincial data, provincial planners need at least municipal data — and municipal planners need data disaggregated to some kind of submunicipal level (*barangay*, *barrio*, village, neighbourhood etc.). If planning is to be pushed down to the *barangay* level, the Data Bank will have to be extended accordingly if it is to give support to the process.

Statistical projections

Various outputs are generated from the Data Bank, including projections, maps and indicators. In view of the different times and sources from which data are drawn, it is particularly important to annotate Data Bank outputs accurately and fully, giving the exact nature of the data being presented as well as the year to which the data refer. Beyond this, some specific comments can be made.

The Ministry has adopted the practice of supplying municipalities with standard projections of certain data for planning purposes. It is felt that this will ensure a degree of consistency and accuracy that might otherwise be hard to obtain. However, in at least one case, this may not be true, namely, municipal population projections. As long as a municipality is growing in a uniform manner, official projections are as reliable as any. However, in cases where a municipality experiences

Table 10 Population forecasts for a hypothetical municipality in the Philippines showing the effect of different forecasting techniques

Population forecasts computed for the year . . .					1985	
Base	1975	Total population			4,000	
Years:	1980	Total population			5,000	
		Average annual increase . .			200	(linear)
		Annual rate of growth			4.46%	(exponential)

	Actual population			*Projected population*				
Barangay	*1975*	*1980*	*Annual growth rate*	*Linear*	*Exponential*	*Share*	*Shift*	*Official forecasts*
Alicia	1,000	1,250	4.46%	1,500	1,563	1,781	1,781	1,563
Bagupan	1,000	1,000	.00%	1,000	1,000	1,000	1,069	1,250
Cumabel	1,000	2,000	13.86%	3,000	4,000	4,125	3,919	2,500
Dungo	1,000	750	-5.75%	500	563	219	356	938
Total	4,000	5,000	4.46%	6,000	7,126	7,125	7,125	6,251
Variance of official forecasts from projected:								
					(Official as a percentage of projected)			
Alicia (average growth)				4.17%	.00%	-12.28%	-12.28%	
Bagupan (no growth)				25.00%	25.00%	25.00%	16.96%	
Cumabel (rapid growth)				-16.67%	-37.50%	-39.39%	-36.20%	
Dungo (negative growth)				87.50%	66.67%	328.57%	163.16%	

Source: The four projection techniques are taken from Neil Sipe and Robert Hopkins, "Microcomputers and Economic Analysis: Spreadsheet Templates for Local Government" (University of Miami, 1984), chapter 4.

Notes:
1. A linear projection assumes each *barangay* grows by the same average annual amount as it did in the past.
2. An exponential projection assumes each *barangay* grows at the same annual rate of exponential growth as it did in the past.
3. A relative-share projection assumes each *barangay* gets the same relative share of total growth as it did in the past.
4. A relative-shift projection assumes each *barangay* shares in total growth relative to a linear projection of its past share.
5. The official forecasts are based on the assumption that each *barangay* grows at the same annual rate of exponential growth as the entire municipality did in the past.
6. The lower part of the table shows the extent to which official forecasts overestimate (positive) or underestimate (negative) the other projections.

areas of relatively high or low growth, official projections become open to question.

As Table 10 illustrates, official projections tend to overestimate the population of areas growing more slowly than the municipal average and to underestimate the population of areas growing more rapidly than the municipal average. (In fact, official projections do not appear to allow for the possibility of negative growth at the submunicipal level.) Because of the exponential nature of population growth, the net effect is to underestimate overall growth in municipalities which are not growing in a uniform manner.

One effect of this is that, in municipalities experiencing urbanization (i.e., population shifts from rural to urban areas), official projections will tend to disguise these trends and to underestimate overall population growth. Similar comments can be made about methods used for making official projections at larger scales (i.e., provincial, regional and national). However, the important point is not that one particular technique is better than others; rather, it is that different techniques are appropriate in different circumstances. No single technique is going to be the best for every municipality in the country. So, instead of trying to provide standardized population projections for every municipality, the Ministry should be trying to encourage municipal planners to select the most appropriate technique and make their own projections based on their detailed understanding of the dynamics of their particular planning area.

Maps

Turning now to the production of maps, there appears to be a tendency to map everything without really thinking through why data are being mapped and what relationships are meant to be revealed by mapping. Consider the question of mapping the number of cattle, hogs and chickens: does it really make sense to map absolute numbers without regard to land area or farm size (if the idea is to show agricultural productivity) or population (if the idea is to show food self-sufficiency)? Does it make sense to map the locational distribution of schools when the important thing is which municipalities that ought to have schools do not have them? It has to be recognized that some information is better not mapped but presented in another form. For example, where the range of data is not very large, differences may be obscured by mapping, particularly if the map is not accompanied by details of the base data.

Indicators

Finally, the Data Bank has been used as the source of "indicators" of the level of services available in different parts of the country. The aim of indicators is to provide a quantitative and objective way of measuring what are normally qualitative and subjective conditions. For example, it may be desired to measure the standard of medical services available in a given area; so a surrogate "indicator" is constructed — according to proposals in a study by the Ministry (Services Delivery Profile, vol. III, p. 27) — by averaging two data:

- the ratio of physicians, nurses and midwives per capita compared to national standards, and
- the ratio of hospital beds and rural health units per capita compared to national standards.

Indicators like these are neither "correct" nor "incorrect" in and of themselves; they are only more or less useful as surrogate measurements of what is the real focus of attention. Moreover, indicators inevitably reflect significant assumptions about what they are trying to measure. In the example cited above, consider some of the assumptions being made:

- that medical services should be uniform throughout the country, regardless of need (structure of the population, birth/mortality rates, incidence of disease, etc.);
- that medical services are a function of resources provided rather than results produced;
- that medical services provided by professionals other than physicians, nurses and midwives or institutions other than hospitals and rural health units are somehow "represented" by those listed;
- that all of the standards (physicians per capita, hospital beds per capita, etc.) are equally important (i.e., there is no weighting);
- that medical services are independent of each other in the sense that adding one more physician to the set of services provided has the same effect regardless of the number of nurses or hospital beds that may be available; and
- that medical services are "substitutable" for each other, in the sense that more of one can make up for less of another.

There may be still broader assumptions implicit in this particular example: for instance, the idea that medical services are in fact a good indicator of health care and, indeed, that public health is a function of curative rather than preventive treatment.

Before adopting any indicator, therefore, it is important to identify and assess the assumptions it reflects. Secondly, it is important to actually test the indicator in practice, to see how far the indicator corresponds to the kind of qualitative and subjective assessment that people would normally make. All of this is especially important if the indicators are themselves to be combined into composite provincial and regional indicators (as the Ministry proposes). Are all the indicators equally important? How, for example, should an indicator of "ecological balance" (based on the number of sanitary toilets per capita) be weighted in relation to an indicator of "sports and recreation" (based on an average of per capita participation in children's sports, playing fields and neighbourhood parks)? Can one indicator compensate for another — a high standard of education services for a low standard of medical services?, or high standards in one municipality for low standards in another? or resources provided for results produced? Questions such as these are implicit in the construction of indicators. Where computerization has removed a lot of the drudgery from calculating indicators, it is all the more important to interpret the results and apply them to planning and decision-making in a careful and judicious way.

Despite these various weaknesses and shortcomings, the Planners' Data Bank represents a considerable achievement. It provides what is probably the best and most consistent view of the state of human settlements in the Philippines. As such, it provides valuable information on general conditions and trends to policy-makers and decision-makers at senior levels of government. Where it does less well is at the detailed planning level.

In fact, because of the two key assumptions discussed earlier — that there is a finite quantity of "important" information and that this can be assembled objectively in one place — the Planners' Data Bank was never really designed to meet the needs of planners. The underlying function of the Data Bank has always been conceived as one of providing a source of basic data — a research or archival kind of role rather than an operational one. By contrast, what planners need is information that is *timely* and *relevant* to specific planning needs and which they can *process* and *manipulate* in a variety of different ways. Neither of these objectives has ever really been a priority for the Planners' Data Bank.

Accordingly, the Ministry should complement its Planners' Data Bank — administration of which may in any case be transferred to the University of Life — with a Human Settlements Information System (HSIS). This Information System would be specifically designed to support human settlements planning and management at a local level as well as to provide programme and management information to senior levels of government. The Information System should be operated by the Ministry of Human Settlements in its capacity as lead agency for town planning within the National Co-ordinating Secretariat.

The key feature of the proposed Human Settlements Information System is that it would be decentralized to operational levels. In this way, it can provide planners at those levels with both data and the capacity to use them. At the same time, use of the system at a local level would provide an automatic mechanism and incentive for regular updating by its users. That in turn would create a source of reliable and current information for management and planning at more senior levels. The Information System would be standardized to the extent that certain common procedures would be followed and certain minimum data requirements would be set. But the system would also be flexible enough for planners to be able to adapt it to special requirements or use it in specialized ways.

Thus, the new Human Settlements Information System would complement the more research-oriented Planners' Data Bank in two specific respects: namely,

- Where the scope and detail of the Data Bank is determined by the need to keep national records of "basic" socio-economic data, the scope and detail of the Information System is governed primarily by the operational needs of planners at the local level, for it is they who will be both users of and suppliers of data to the system.
- Where use of the Data Bank is intended mainly to produce standardized tables and maps, use of the Information System is designed to give planners at the local level access to their databases in electronic form, so that analysis and presentation of information becomes an element in the planning process.

Moreover, the database for such an Information System cannot be prescribed in advance. It has to be built "from the ground up" out of the practical, operational concerns of planners. That means going out into the regions and provinces and identifying:

- Who specifically are to be the "clients" for the Information System and what kinds of decisions do they make?
- What specific kinds of information do they need in order to be able to do their job?
- What data are already available and in use because they are known to be up to date and reliable?

The objective of such research is to select key data for critical planning issues rather than basic data for planning in general. Some of the "key data" may vary from place to place, as issues that are critical in one part of the country may not be so important in others. On the other hand, data on land use or municipal finance are likely to be critical to planning almost everywhere. It is also important to identify planners' current data sources — electoral rolls, school enrolment records, water and electricity accounts, licence registries, etc. — in order to see how far such sources can supplement formalized methods of data gathering such as censuses, surveys and questionnaires. Not only are the former a lot cheaper than the latter, but the former are usually kept up to date on a regular basis in the ordinary course of their operational use.

At the same time, a similar inquiry should be made at senior levels in order to identify specific clients and their information needs. Once again, the objective is to select key data requirements in terms of precise decision-making needs. On that basis, a set of common standards and procedures can be developed for the Information System. This will ensure that relevant data can be abstracted from local versions of the Information System and aggregated to regional and national levels to provide the information required.

The important point about proceeding incrementally and empirically and designing the Information System on the basis of demonstrated needs rather than abstract principles is that it secures commitment to use and updating of the system. The result will be something less than a single comprehensive database, but, if managed properly, the system will be a coherent "system of systems" — some large and some small, some local and some regional, some frankly better than others — but all of them consistent with one another to a degree sufficient for data-sharing yet flexible enough to be able to meet the needs of their primary users. In this way, each system can serve both local and central planning needs.

The technological key to such an Information System is the microcomputer. Thanks to the development of these low-cost, "user-friendly" machines and their massive libraries of software "packages", it is now feasible from the point of view of cost and training to give serious consideration to a decentralized network of users along the lines described above. If the equipment is compatible, data can be shared by the simple expedient of exchanging floppy disks. Although it might seem attractive, on-line electronic communication is hardly necessary for human settlements planning, since little of the pertinent data will decay in the few days it may take to send a floppy disk from one part of the country to another. If data are maintained in standard data files, they can be read and processed using standard programmes for word processing, database management, spreadsheet analysis, accounting, statistics and even mapping.

The Ministry already has a microcomputer in each of its 12 Regional Offices; the next logical target would be the municipalities. There are anywhere from 60 to 100 or so of these, depending on how the definition is drawn. Almost all of these have an identifiable planning unit in their administration, which could take charge of a microcomputer system and participate in a training programme. Most also have an approved town plan and zoning ordinance which they are implementing, along with appropriate capital spending programmes. These municipal planning agencies are the obvious participants in and contributors to a Human Settlements Information System.

Case 12

A land management information system (Mauritius)

The Ministry of Housing, Lands and the Environment is the governmental agency chiefly responsible for human settlements planning and management in Mauritius. It is divided into three main units: the Survey Division, the Planning Division and the Administration and Finance Division. The Ministry is also responsible for two important parastatal agencies. One is the Town and Country Planning Board (TCPB), which advises local authorities on matters connected with planning under the Town and Country Planning Act. The other is the Central Housing Authority (CHA), which is responsible for building and financing public housing for victims of cyclone damage and currently administers some 25,000 units.

Neither the Ministry nor its two parastatals are computerized. In November 1986, following the recommendation of its 1984 consultant, US AID provided one microcomputer to the Ministry. The computer is an IBM PC-AT with 512 Kb RAM, standard colour-graphics monitor and 20 Mb hard disk; the peripheral equipment consists of an Epson LQ-1500 dot-matrix printer, an Epson plotter and a Summagraphics mouse. No applications software was supplied with the computer, although Ministry staff appear to have been able to obtain some programmes. No other equipment is available in the Ministry.

One of the most important points to recognize at the outset is that the practical limit on the use of a computer is often not the capacity of its RAM (memory) or the speed of its CPU (central processing unit), but rather user access to it. For example:

- It may seem an obvious point, but only one person can use a keyboard or screen at one time; meanwhile, other would-be users have to wait their turn, even if the computer itself is doing nothing more complicated than word processing or accounting. If there are 40 hours in the work-week, the total number of person-hours available on the computer is only 40. (It is worth noting that a flexible approach to working hours (e.g., letting staff work for any eight-hour period between, say, eight a.m. and six p.m. daily instead of only for a fixed eight-hour period) would add 25 per cent to available computer time — quite apart from other social and economic benefits that flexible work-hours might bring.)

- Secondly, a computer has to be easy to get to and easy to use if people are to be persuaded to abandon the manual ways they are familiar with — no matter how attractive the promised benefits of computerization may appear. Thus, the idea that computers should be locked away in special rooms and access to them tightly controlled should be replaced by the principle of making access as easy and casual as is consistent with security and shared use.
- Thirdly, training and support should be easily available to all. Quite a lot can be done to hide from the ordinary user the complexities of repetitive, standardized computer applications; but every effort should be made to encourage computer literacy and self-improvement among those motivated enough to ask for it.
- Fourthly, while it is perfectly possible for basic microcomputers to process in seconds databases that contain millions of characters, those data must first be typed into the computer and then kept up to date on a regular basis. This can take a great deal of time, both staff time and computer time, particularly if data-entry conditions are not ideal — if, for example, data sources are poorly organized or physically awkward to handle. Thus, while a reasonable copy-typist can manage about 15 Kb of text per hour (or, say, half a megabyte a week), data-entry speeds for ordinary databases are typically only 10 or 20 per cent of these figures. Database maintenance can also be very time-consuming if it is not properly and systematically organized.

Another important consideration is the need for an incremental approach to computerization. According to this view, progress is achieved by means of small, practical improvements rather than in accordance with some grand design worked out in advance. There are several practical reasons for adopting this kind of approach:

- First, there are tactical advantages to getting small-scale computer applications into operation as soon as possible, in order to give tangible evidence of potential benefits as well as to provide users with hands-on experience of both the process and the pay-off of computerization.
- Secondly, small-scale applications can always be integrated into large ones in due course. Naturally, it is desirable to bear this in mind when designing the original applications, in order to make eventual integration as easy as possible.

- Thirdly, even when computers are available, some things are still best done by hand. (Just as with the availability of the telephone, some conversations are still best held face-to-face.) The optimal data environment for an organization will probably turn out to be a hybrid of various systems, some computerized and some not.

Even when it comes to hardware, the old idea that you should try to anticipate future requirements as far as possible and design your hardware accordingly has become much less important in the past few years. For one thing, the modularization of equipment that has accompanied the microcomputer revolution makes it possible to "mix and match" equipment from different suppliers to a far greater extent than used to be the case. Moreover, microcomputer equipment tends to adhere to standards that, even in the case of the so-called "IBM compatible" standard, are no longer the preserve of a single company; users are no longer so much at the mercy of a company that decides to phase out a particular line of computers. Of course, this does not mean you can buy equipment with no thought to long-term needs, to compatibility with future equipment, to potential reallocation to low-priority uses, etc.; but it does mean that it is quite feasible now to proceed with computerization based on gradually acquiring hardware over time rather than making a commitment to a single manufacturer or even a single type of system at the outset.

Indeed, even from the hardware point of view, there is much to be said for proceeding incrementally.

- First, the risk of outgrowing the capabilities of computer hardware is ever-present, no matter what initial configuration is bought. The trick is thus to try to determine how soon this is likely to happen for a given configuration and how much extra capacity that warrants buying. For with computer technology advancing as rapidly as it has in the past few years, the sensible thing is *not* to buy any hardware until it is actually needed.
- Secondly, learning to use basic equipment to the full extent of its capabilities provides staff with a much better basis than they initially have on which to judge what additional capabilities they require. It is always difficult to assess computer needs in advance for the simple reason that supply tends to create demand: it is not until you have access to a computer that you can really begin to see what you can do with it or even what are your real priorities.
- Thirdly, it is not at all undesirable that an organization should grow by means of incremental investment in capital equipment rather than by an attempt to meet all its potential needs at the outset. In fact, that is how most capital spending decisions are

made; computers are no different. If and when needs do outgrow current capacity (as long as it does not happen too soon after the original purchase) that does not mean a mistake was made. On the contrary, it suggests that existing equipment has been thoroughly used and that (had it been available) additional capacity would have gone unused all this time.

- Finally, there is nothing to lose by starting out with a simple computer system; it can always be upgraded to an elaborate one, if and when the need arises and the resources become available. In spite of the rate of technological development, many of today's microcomputers will be able to serve for years to come as perfectly useful word processors and/or as terminals to large computer systems; capital investment in microcomputers is not likely to be wasted even if some users eventually do outgrow capacities. Similarly, time and money spent entering data into a computer are not likely to be wasted even if the hardware and software are changed. Although not all computers and programmes are compatible with one another, there is almost always a way of transferring data files from one system to another.

In summary, the Ministry of Housing, Lands and the Environment is at a point where an increasing number of organizations all over the world now find themselves. Now that computerization is manifestly within their grasp, financially and technically, in a way that it never was before — what is to be done? Indeed the possibilities are suddenly so numerous and exciting that it can be difficult to avoid paralysis on the one hand or unrestrained enthusiasm on the other. For just as certainly as there are potential benefits of computerization, so too are there risks. As with any other organizational decision, computerization warrants careful planning if benefits are to be maximized and risks minimized.

One of the priority uses identified for the new microcomputer is computerization of Crown Lands and Leases. These include all lands held or reacquired by the government, including the *Pas Géometriques*, which was originally an 81-metre strip of land along most of the foreshore established by the French authorities for defence purposes. (Eighty-one metres is approximately 89 French feet, but why that particular figure was chosen is not clear (Meade, p. 6).)

To help keep track of all the deeds, leases, plans and other legal documents pertaining to Crown Lands, the Survey Division has two important manual databases: the *Domaine Book* and the *Index of Leases*. The first is a single copy of a published inventory of all Crown Lands in Mauritius as of the year 1934, which has been kept up to date by means of hand-written annotations. The second is a chronological register of all leases and vestments involving Crown Lands. The documents referred to in both the *Domaine Book* and the *Index of Leases* are archived within the Ministry's own premises. Both databases are organized on the basis of the nine colonial Districts. (Under the colonial administration, the nine Districts were: Port Louis; Plaines Wilhems (now divided among the other four municipal authorities and DC Moka-Flacq); Moka and Flacq; Black River (now divided between DC Moka-Flacq and DC South); Savanne and Grand Port (now part of DC South); and Pamplemousses and Rivière du Rempart (now DC North).)

Information in the *Domaine Book* is complemented by a series of Land Acquisition and Sales Books — the latter quite compact, since Crown Lands are not often disposed of — which are meant to chronicle all transactions in Crown Land. These Books contain substantially the same information on each Lot as the *Domaine Book*, again on the basis of the old Districts. However, according to the 1985 Ramkissoon Fact-Finding Committee on Crown Lands (Part I, pp. 89-91), there are significant disparities between these Books and the *Domaine Book* in both the period before publication of the latter and in the years since.

Similarly, the Ministry's archives are complemented by those kept by the Office of the Registrar-General, who is responsible for registering all deeds and leases in Mauritius, including those pertaining to Crown Land. No detailed comparison of records in the Registrar-General's possession with those of the Survey Division has apparently been undertaken up to now.

There are at least three important reasons for proceeding as quickly as possible with computerization of the Ministry's records of Crown Land. These are as follows:

1. Security

The manual databases now in use are unique. The consequences of their loss or destruction (e.g., by fire) are beyond calculation. Yet no special precautions are being taken to safeguard these documents, nor apparently has it been feasible to maintain duplicate copies of either the *Domaine Book* or the *Index of Leases*.

To make matters worse, these records are daily subject to gradual destruction through constant handling — to the extent that the 1985 Fact-Finding Committee (Part I, p. 89) described the current state of the *Domaine Book* as "pitiful" and warned that its contents would soon become "unintelligible and irretrievable" if changes were not soon made in existing procedures.

Computerization would mean that (a) backup copies of all the databases could be kept in different places, both as a safeguard against accidental loss and to facilitate access at different locations; and (b) up-to-date printed copies could be produced whenever they are required, either because previous copies have become worn out or because they have just got too difficult to read owing to amendments.

2. Analysis and checking

The manual databases now in use are essentially unusable for any analytical purposes. The 1985 Fact-Finding Committee (*loc. cit.*) describes the present system as "so unmanageable" that it precludes anyone from "undertaking a comprehensive survey or any other meaningful analysis of Crown Lands". "As a result", the Committee continues, "no *Comprehensive Survey* of Crown Lands has been attempted for the past half-century. . . . And there is no doubt that it is *such absence* that has been the *root cause* of most of the administrative problems mentioned in this Report" (Part I, p. 98).

Computerization would mean both that existing information could be processed efficiently and that systematic efforts could be made to find and correct erroneous and out-of-date information. For example, it would take only a matter of seconds to determine (subject to the accuracy of the data in the database) how much Crown Land there is, how much of it is leased, what leases expire in the next year and so on.

Similarly, if lot extents calculated in old surveys are now regarded as suspect (*vide op. cit.*, p. 88), it would be a simple matter for the computer to list all leased Crown Land that has not been surveyed within the past (say) 50 years and where the extent of land shown in the *Domaine Book* does not correspond to the extent shown as leased; these particular cases could then be checked in the field and verified accordingly.

3. Administration

It is estimated that approximately 7,000 of the 10,000 lots of Crown Land are currently subject to vestment or lease. The latter alone entail some 12,000 different leases or leasehold agreements, each with its own terms, conditions and date of expiry.

In the case of leases, not the least important consideration is the timely collection of rents which is undertaken by the Finance Division on the basis of its own set of records. According to Ministry staff, nearly 20 per cent of Crown Land lease accounts are currently in arrears. Total arrears as at June 30th, 1986 was about MRs 1.5 million, of which more than half is more than 12 months in arrears. Additional revenue losses are caused by delays in registering new leases and delays in renewing old leases.

Computerization would make the administration of vested and leased land effective and efficient. Not only would it mean that key information was readily available — for example, what leases expire this year? which leases are in arrears? and how much money is involved? — but it would also be possible for a computer to prepare and address notices of various kinds, such as reminders of arrears, notices about the expiry of leases and survey questionnaires about current land use or current stage of project completion. Under the present manual system, for example, it appears that notification of arrears occurring on 1 July does not take place until October or November and no further action is taken until February or March. In part, this is because existing staff are unable to deal with the number of records and the quantity of paperwork involved; computers are very good at preparing large numbers of form letters.

In view of the foregoing, there is a clear and compelling case for proceeding with computerization of the Ministry's Crown Land records as a matter of high priority.

It is further proposed that computerization should follow fairly closely the manual system operated by the Survey Division. The reason for this recommendation is not that the present system is ideal or that its information is always accurate and essential, for this may not necessarily be the case. Instead, there are three other reasons for starting out as closely as practicable to the system already in place and in operation:

- First, there is already enough uncertainty about where to get the most accurate and up-to-date information on Crown Lands for it to be unwise to create yet another database (albeit computerized) that is not clearly and closely related to existing ones.
- Secondly, data-entry will be easiest if data can be keyed into the computer directly from existing records; thus, fields should preferably be in the same form and sequence in the computerized database as they are in the manual ones, at least until data-entry is completed, after which fields can be rearranged and/or added to if desired.
- Thirdly, even if several different units (including the Planning and Finance Divisions) are able to make effective use of the databases once they are computerized, this will still depend (as it does now) on the Survey Division keeping the databases up to date; so, in order to ensure the commitment of the Survey Division, it is vital for that Division to feel that its own needs (as well as those of others) will be met better by the computerized databases than by the manual ones.

Table 11 Crown Lands: Domaine Database (Mauritius)

Name	*Size/Type*	*Description of Field*
DISNAM	2 C	District name (PL, PM, RR, FL, GP, SA, BR, PW, MO)
DOMLOT	5 C	Domaine Lot no. (including Crown Lot no. extension)
DESCRI	30 C	Description of lot in plain language
EXTENT	8 N	Area of lot in arpents (to two decimal places)
ORIGIN	1 C	Code for origin of land (U=unconceded; A=acquired)
ACQDAT	8 N	Date acquired (YYYYMMDD)
ACQNAM	25 C*	Name of vendor (note 3)
ACQPRI	7 N	Price when acquired
TRVNUM	6 C	Transcription-Ventes number of deed of purchase
SOANUM	6 C	Survey Office Archive number (if applicable)
SURNAM	10 C*	Name of surveyor (note 3)
SURTYP	1 C	Code for type of survey (memo or plan)
SURDAT	8 N	Date of survey (YYYYMMDD)
SURLOC	6 C	Reference number for storage location of survey
SURSIG	1 C	Survey signed by neighbours (Y/N)
PERIME	8 N	Perimeter of lot when surveyed (in metres)
MAPNUM	4 N	Map sheet number at 1:2,500 scale (if available)
MAPREF	4 N	Grid reference on MAPNUM
LOCALE	3 N	CSO code for exact locality where lot is situated
Total	143	Approximate file size for 10,000 records = 1.40 Mb

Notes:

1. Fields are designed and arranged with a view to facilitating data-entry; they can be rearranged later if desired. Under File Type, "N" indicates fields restricted to numerical data, whereas "C" indicates fields suitable for any character or alphanumeric data.
2. Field sizes are only estimated. Unless an abbreviation is universally accepted (e.g., St for Saint), data in fields marked with an asterisk (*) should be truncated rather than abbreviated to fit the available space. Full stops and other unnecessary punctuation should be omitted.
3. Proper names should be entered surname first, first name/initials second; similarly government departments should be entered using the form, "Planning, Ministry of".
4. If a check of the entire database shows that the date of acquisition of all acquired land is known, then unconceded land can be coded as 00000000 (or similar) and the ORIGIN field can be dispensed with.

Thus, it is recommended that the computerized databases should be made to look and feel as much like the manual ones as possible, at least in the beginning. Detailed recommendations for the structure of the two databases are provided in Tables 11 and 12. The first is meant to duplicate the *Domaine Book* and the second the *Index of Leases*, although in fact one item has been omitted from and one group of items has been added to each database.

The only field omitted from both databases is the "Remarks" column. This decision was taken not on the grounds that the information in that column is somehow less important than other kinds of information, but rather because remarks and comments do not need to be computerized since they rarely need "computerizing". Information in the "Remarks" column can just as well be stored in hardcopy form (i.e., on paper, in a book or on microfiche).

Three fields have been added to the Domaine Database (Table 11) to help in identifying the exact location of a plot. First, each record contains the four-digit number of the relevant 1:2,500 map sheet plus a four-digit grid reference to it; the effect is to create an eight-digit "geocode" that is accurate enough to locate a plot to within a ten-metre square. Second, there is a field for specifying the location of each plot using the three-digit code developed by the Central Statistical Office for some 800 different localities in Mauritius.

In the case of the Lease Database (Table 12), extensive additions have been made primarily to assist the Finance Division with the collection of rents. Essentially the question was whether and, if so, how to make the computerized database useful for administering as well as recording leases. Eventually, of course, it would be desirable to computerize the entire process of managing rent accounts, including on-line transactions and regular statements of account. However, for the present, neither the volume of transactions nor the available computer capacity justifies such a step. Instead, what is proposed here is the addition of several fields to the Lease Database. These additions will not be sufficient to replace the existing manual system of accounts; but they will do a lot to help collect rents and monitor arrears effectively and they will provide a useful check of the existing manual system.

Certain of the information required for lease management is already in the Lease Database, e.g., the name of the lessee (LESNAM), the date of commencement of the lease (COMDAT), the date of expiry of the lease (EXPDAT) and the current annual rent (CURREN). What needs to be added is the current address of the lessee and critical account-balance information. Thus, the following five fields have been added to the Lease Database:

Table 12 Crown Lands: Lease Database (Mauritius)

Name	*Size/Type*	*Description of Field*
DISNAM	2 C	District name (PL, PM, RR, FL, GP, SA, BR, PW, MO)
DOMLOT	5 C	Domaine Lot no. (including Crown Lot no. extension)
LOTNUM	6 C	Building Lot number (if applicable)
CLANUM	3 N	Land classification number (if available)
STATUS	1 C	Code for disposition of land (L=leased; V=vested)
LESNAM	20 C*	Name of lessee/vestee (note 3)
LESEXT	8 N	Extent of land subject to lease/vestment (note 4)
LESNUM	5 C	Lease Book volume no. (Transcription-Baux number)
LESTYP	1 C	Code for type of lease (note 5)
COMDAT	6 N	Commencement date of lease/vestment (YYMMDD)
EXPDAT	6 N	Expiry date of lease/vestment (YYMMDD)
CURREN	6 N	Current annual rent payment
FILNUM	10 C	Ministry file number
PURPOS	2 C	Code for stated purpose of land (note 5)
RENDAT	2 N	Last year when current rent (CURRENT) applicable
CURBAL	9 N	Current outstanding balance as per manual accounts
CURDAT	6 N	Date of current outstanding balance (CURBAL)
LESAD1	20 C*	Street/box address of lessee/vestee
LESAD2	15 C*	Town address of lessee/vestee
Total	133	Approximate file size for 12,000 records = 1.55 Mb

Notes:

1. Fields are designed and arranged with a view to facilitating data-entry; they can be rearranged later if desired. Under File Type, "N" indicates fields restricted to numerical data, whereas "C" indicates fields suitable for any character or alphanumeric data.
2. Field sizes are only estimated. Unless an abbreviation is universally accepted (e.g., St for Saint), data in fields marked with an asterisk (*) should be truncated rather than abbreviated to fit the available space. Full stops and other unnecessary punctuation should be omitted.
3. Proper names should be entered surname first, first name/initials second; similarly government departments should be entered using the form, "Planning, Ministry of".
4. Once data-entry and checking are complete, it would be desirable if the LESEXT field in this database could be reconciled with the corresponding EXTENT field in the Domaine Database and the LESEXT field deleted from the present database.
5. Appropriate codes for LESTYP and PURPOS are not currently in use and will accordingly need to be developed and applied to data in the *Index of Leases*.

LESAD1 and LESAD2 to record the lessee's current postal address;
RENDAT to show the last year in which the current annual rent (CURREN) is applicable before it has to be adjusted under the terms of the lease; and
CURBAL and CURDAT to show the current account balance (i.e., arrears), including principal and interest, and the date it was (last) calculated.

The purpose of this additional information is to enable the computer to do such things as:

(a) send notices of arrears, rent adjustments, lease renewal dates, etc. to lessees at the appropriate times;
(b) list all arrears or analyse arrears by location, type of lease, size of plot, etc.; and
(c) identify accounts where no payments have been made in (say) the past six months or where the amount outstanding is more than a certain amount.

Once the databases are in electronic form and staff in all Divisions have had some experience in using them, substantial kinds of innovation can be considered.

- The Survey Division can start to check its own data on a systematic basis, possibly following some of the suggestions contained in the report of the 1985 Fact-Finding Committee (e.g., reconciling its own different sources, checking existing data, finding missing data for existing records, adding records that are entirely missing, collecting information from other Ministries, etc.).
- Another possibility is to add a field for extent in hectares and then, in a matter of seconds, have the computer calculate the extent of all lots in metric units instead of or as well as *arpents*. (One *arpent* is approximately 0.4221 ha, compared to an acre which is approximately 0.4047 ha). It goes without saying that all such calculations would be done without any errors of arithmetic!
- Similarly, a field for local authority (AUTHOR) could be added and lots coded according to the local authority where they are located as well as or instead of the now somewhat archaic colonial district.
- It may also prove possible to consolidate the computerized version of the databases and even to discard some fields as redundant or unnecessary. It is important to remember that computers should be used for storing information that is subject to change or information that needs to be sorted, searched or otherwise processed. There is little point in putting into a computer information that

never changes or is never used, just for the sake of storing it there; hardcopy (such as paper or microfiche) is the best medium for storing information that is just being kept "for the record".

- The Survey Division may decide to add new fields for data on land characteristics such as soil type, elevation or slope. However, it should be remembered that, if data in a database are subject to periodic change (land use as opposed to, say, soil type), such data are going to need periodic updating — and that means finding someone or some agency able and willing to do it on a regular basis.
- Other users (such as the Planning and Finance Divisions) can extract from the Survey Division's "master" databases those fields which they need. For example, a database such as the one described by N. Patten in the appendix to his report on his UNCHS-sponsored training programme could be constructed by taking selected fields from each of the databases proposed. (Note that databases extracted from other databases in this way can easily be updated from the original databases whenever it is desired.)
- Similarly, these other users may thereafter be in a position to contribute new data fields to the databases maintained by the Survey Division. For example, if the databases are to contain information on "development potential" or, even, on current land use, it may well be the Planning Division rather than the Survey Division that is in the best position to provide such data. Similarly, data on leasehold arrears probably has to come from the Finance Division.
- Finally, there is the possibility of integrating with the Ministry's Crown Lands databases information on Crown Lands administered by other agencies of government, information on Crown Lands maintained by the Registrar-General, information on land owned by parastatals and local authorities and so on. Thus, it is quite reasonable to hope that these modest steps towards computerization of the existing Ministry databases will lead in due course to a comprehensive information system on publicly owned land, such as was proposed by the 1985 Fact-Finding Committee (Part I, chapter 11, section 4 *et passim*).

In order to get these databases into operation, quite a lot of information will have to be entered into the computer at the outset — an estimated 10,000 individual lot records in the case of the Domaine Database and about 12,000 lease or lease-agreement records in the case of the Lease Database. In the case of the Domaine Database, responsibility will fall primarily on the Survey Division to ensure that all the relevant information from the *Domaine Book* is entered and checked. The computer can then be used to generate lists of missing information such as Map Sheet numbers and grid reference numbers which will have to be provided from another source. If it takes an average of perhaps two minutes to enter each listing from the *Domaine Book*, it will take about 10 weeks to enter the entire book plus two more weeks for verification and adding further information. In the case of the Lease Database, initial data-entry will again have to come from the Survey Division. Then the Finance Division can add data from its records. Next a list of disparities between the two sets of records can be printed and an attempt made to resolve them. For example, the Survey Division estimates its records include 7,000 leases, whereas the Finance Division estimates that it administers 12,000 leases or lease agreements. Using the larger figure, if it takes an average of two minutes to enter the details of each lease, it will take about 15 weeks to accomplish the basic data-entry plus five more weeks for verification, for addition of information from the Finance Division files and for resolving anomalies.

Once initial data-entry is completed, maintenance will be minimal in the case of the Domaine Database, since all that is required is to add in new purchases and sales of Crown Land. In the case of the Lease Database, however, maintenance is a significant activity. Every time a payment is made on a lease account, for example, the corresponding record in the Lease Database will have to be retrieved and edited. If each account is updated twice per year, that means an average of 500 account adjustments each week; if each account takes 30 seconds to retrieve and edit, the process will take an average of four hours per week of computer time. Similarly, if preparing notices of renewals (say 1,200 per annum or 25 per week) and notices of arrears (say 2,400 per annum or 50 per week) are assumed to take a minute each, then that will mean another couple of hours each week. In short, maintenance of the two Crown Lands databases is likely to require a minimum of a full day of computer time every week.

In summary, with its Crown Lands and Leases database system, the Ministry has the opportunity to make an immediate and worthwhile beginning to computerization. The resources are there; all that is required is the commitment to proceed. As the Titmuss Report concluded more than a quarter of a century ago,

> It is the task of the outside adviser to sift the facts, analyse the problems and present alternatives to [government officials]. It is they who have to shoulder the responsibility of choosing a course of action. . . . To be effective, both require vision; a quality which we are confident is not lacking in the people of Mauritius.
>
> (Titmuss Report, p. 248)

Part four

Institutional factors

Part four

Institutional factors

Computer systems, it goes without saying, do not have an independent existence. They exist in and for the benefit of organizations, where the computer is in constant interaction with the people and structures which make up the organization. Thus, the design of a computer system should always reflect basic institutional realities: who is going to use the system? what are they going to use it for? and how do they propose to use it? The relationship between institutions and computers works the other way as well. Just as organizations shape the design of a computer system, so too the computer system shapes the organization — by increasing its efficiency, by expanding its knowledge base, by adding to the power and prestige of certain of its units, by readjusting the division of labour between centre and periphery and so on. These factors also need to be taken into account in the design of computer systems.

This section presents five cases. The first two focus on the place of leadership in identifying user needs and setting objectives for computerization. The first case (from Bahrain) deals with a situation where, owing to their lack of experience with computers, users will need encouragement and assistance in defining their needs. The second case (from Turkey) deals with a situation where user needs could very quickly overwhelm available resources, making leadership essential for setting priorities and avoiding fragmentation of effort.

The other three cases discuss various aspects of the complex relationship between computers and organizations. One case (from Swaziland) describes a human settlements Ministry faced with ever-growing demands for its services at a time when the government has decided to freeze all staff hiring. Computerization offers a promising opportunity to improve performance by increasing the productivity of existing staff. Another case (from Jordan) examines the opposite kind of situation — one in which a successfully implemented computer system has failed to contribute as much as expected to organizational performance or staff productivity. The last case (from Sri Lanka) discusses how, if carefully implemented, computerization can help strengthen institutional capacity among local government authorities.

The lesson of all these cases is that computerization is not an end in itself but only a means to an end. Computer systems are neither good nor bad except in terms of how they are used. The most elaborate and sophisticated system is worth nothing if it is not used. On the other hand, a simple and straightforward system that really does help people do their jobs is much to be desired.

Case 13: Identifying user needs (Bahrain)

In Bahrain, the Physical Planning Directorate (PPD) of the Ministry of Housing has found itself falling behind in terms of computerization relative to other directorates in its own Ministry as well as other governmental agencies with which it works. Accordingly, PPD determined to begin by computerizing some of its own work, specifically its data management activities.

The case begins by identifying four critical and interdependent aspects to computerized data management. These are:

- collecting only data that are genuinely useful;
- keeping the data up to date;
- making them accessible; and
- teaching people how to use the data.

Ignore any one of these factors and the entire information system will suffer.

Next, the case discusses a number of key points related to the design and implementation of computer systems. First, there is the role of a "lead agent" — a person or group of people willing and able to provide leadership during the process. Secondly, there is the importance of recognizing how much more computers can do with data than just collecting, storing and retrieving them. Examples are provided showing how computers can be used for sharing, manipulating and modelling data in ways that are just not practicable in a manual environment. Finally, the case discusses the critical issue of accessibility and identifies four distinct dimensions to it: the number and physical location of the computer equipment, its ease of use, the availability of appropriate software and access to training and other forms of support. All of these factors are essential to the success of computerization.

Accordingly, the case recommends that PPD should formally designate one of its Sections as a "lead agent" for computerization. Also, in order to maximize accessibility, PPD should buy as many "basic" machines as its budget will allow. This may mean relegating the purchase of specialized peripherals and sophisticated equipment to a later stage. By that time, however, staff members will have acquired skills and have a good idea of how they want to use computers. Similarly, the report recommends use of off-the-shelf rather than custom-written software and careful attention to providing staff training and support from the moment equipment arrives.

Case 14: Setting priorities (Turkey)

In Turkey, the Ministry of Public Works and Settlement is responsible for carrying out civil works and repairs in respect of public buildings and facilities and for certain activities in respect of physical planning and disaster relief. The Ministry's total staff is about 80,000. The Computer Centre in the Data Processing Department is equipped with a Data General MV-4000 Eclipse connected to 15 terminals and three microcomputers. In the short term, the computer is to be used for four main applications: monitoring the disbursement of municipal grants, keeping a roster of civil contractors, managing personnel records and payroll for the Ministry and keeping a scientific and technical database on construction and building materials used in Turkey.

The case presented here assesses the many and varied potential computer applications in the Ministry and concludes that the most critical issue facing the Computer Centre is going to be co-ordinating all these applications without having either the power or the duty to control them. To this end, the report recommends that the Computer Centre adopt a policy of balanced development of both centralized/minicomputer systems and decentralized/microcomputer systems. In conclusion, the report calls for development of an explicit computerization plan for the Ministry as a whole, based on balanced development of computer resources, a leading role for the Computer Centre and in-house staff support for computerization.

Case 15: Increasing productivity (Swaziland)

In Swaziland, the Ministry of Natural Resources, Land Utilization and Energy is the chief governmental agency responsible for human settlements. Six of its departments and divisions are directly involved with human settlements management, namely, the Physical Planning Branch, the Land Use Planning Section, the Housing Branch, the Survey Branch, the Deeds Registry and the Lands Branch. In effect, the Ministry acts as a national physical-planning agency for Swaziland.

Two factors currently dominate planning in the Ministry. On the one hand, it is recognized that demand for its services is growing and will continue to grow as long as population continues to grow and urban migration continues to occur. On the other hand, the Ministry has been advised not to expect any increase in its operating budget during the five years of the current Plan. Thus, it is clear that any improvement in performance by the Ministry is going to have to come not from increases in staff but from increases in productivity.

In several of its branches, the Ministry has been making limited use of microcomputers as a means of increasing staff productivity. Most applications have been fairly technical in nature (e.g., verifying field survey calculations). This case discusses some of the opportunities for expanding computer use to a wide range of Ministry functions, but also warns of the dangers of trying to go too far too fast. When it comes to computers, it is important not only that they work but also that they be seen to work. Thus, the first priority is to get computer systems operational and contributing to staff productivity. Elaborate applications (the case argues) can be developed and implemented in due course, once the Ministry has managed to assemble technical experience and political support.

Case 16: Improving computer use (Jordan)

In Jordan, the Housing Corporation was established in 1966 to provide housing for low-income and middle-income Jordanians in accordance with national objectives. In 1984, based on advice from governmental computer specialists, the Corporation bought a DEC PDP-11/23 minicomputer and two custom-written software programmes, one for monitoring loan repayments and the other for handling the general finance and accounting requirements of the Corporation.

The report deals with its mandate from three points of view: how far the present system meets the defined needs of the Corporation; whether additional computer applications could be identified and met by the present computer system; and what steps could be taken to promote use of computers generally in the Corporation. The report concludes by proposing four sequential phases of activity.

> The first step is to raise computer consciousness in the Corporation by a series of simple steps that cost little or nothing and can begin immediately.
>
> The next step is to institutionalize corporate commitment to computerization by creating a full-time computer staff position and establishing an explicit plan for computer development in the Corporation.

The third step is to enhance the capabilities of the existing computer system, especially so as to make it easily accessible to potential users throughout the Corporation.

Finally, the Corporation should prepare a long-term strategy for development of future computer systems and capabilities.

Case 17: Building institutional capacity (Sri Lanka)

In Sri Lanka, the Ministry of Local Government, Housing and Construction is responsible for overseeing the functions of urban local authorities in the island. There is nearly unanimous agreement on the fact that local authorities have become less and less able to carry out a more and more limited range of functions. Thus, one of the priorities for the Ministry is to strengthen the institutional capacities of the local authorities, and one of the key proposals for achieving this objective has been the establishment of a management information system to monitor the progress of development projects at the local level.

The case examines the proposed management information system and concludes that it could be improved both in detail and in concept. In terms of detailed design, the information system is not sufficiently disaggregated to be able to provide useful information to local authorities; its focus is sectoral rather than project-oriented; it does not distinguish adequately between base standards ("stocks") and changes over time ("flows"); finally, the system provides only for annual data and cannot accommodate monthly or even quarterly figures.

At the same time, the fundamental concept of the proposed information system is too centralized, since the system is designed primarily to enhance central control over local authorities. For one thing, there is little indication that central control will lead to a strengthening of the capacities of local authorities; in the past, central control has often had the opposite effect. For another thing, centralized systems are unlikely to elicit the support of users in the local authorities — support that will be critical to maintaining and updating the system in the future.

Consequently, the case endorses the principle of using computers as a means of strengthening institutional capacity but emphasises that this capacity has to be designed into the application if it is to have the desired effect. In particular, the case recommends that the proposed management information system be designed, first, to give direct support to local authorities and, only secondly, to serve the needs of central monitoring and co-ordination. Provided the system is consistent from one local authority to another, it will be easy to arrange for data from each local authority to be forwarded to the Ministry in Colombo for consolidated analysis on a regular basis — even if the data have to be mailed in the post! The critical factor will be the accuracy and timeliness of the data, and the best way to guarantee that is not by compelling local authorities to meet the needs of the Ministry but by encouraging local authorities to get and use accurate and timely data for their own purposes.

Case 13

Identifying user needs (Bahrain)

The mandate of the Ministry of Housing in Bahrain and that of the Physical Planning Directorate in particular extends to the planning of housing — indeed, of human settlements — in the broadest sense of the term. Effectively, the Ministry is the lead public agency in respect of national physical planning in Bahrain.

At the same time, the Ministry of Housing is critically dependent on a variety of other agencies as sources of data for and as actors in the planning process. Among these other planning-related agencies are the following:

The Ministry of Works, Power and Water, which is responsible for the planning, construction and maintenance of almost all physical infrastructure throughout the country, including roads, street-lighting, sewerage, water, drainage and electrical systems.

Various other governmental agencies, including the Ministries of Health, Education, Justice and Islamic Affairs (which maintains the national Land Register), Transportation (which provides public transport services), Labour and Social Affairs, Commerce and Agriculture (including fisheries and harbours), Development and Industry; the Central Statistical Office; the Bahrain Telecommunications Company (BATELCO); and the Housing Bank.

Municipal governments, through which development control is ultimately exercised by means of their authority to issue building permits.

Together these agencies have been responsible for an impressive list of accomplishments over the past two decades. Two planned new towns have been built: Isa Town, begun in 1963, and Madimat Hamad, begun in 1980. Thousands of housing units have been constructed and assigned to qualified families. Extensive land reclamation has been undertaken at numerous points around the northern end of the island (from Al Budayyi in the west to Sitra in the east), including recently an area of 250 ha near Sanabis, which will eventually provide living space for some 60,000 people.

To support these and other development priorities in Bahrain, the government is already heavily committed to improved information management in public administration. Since 1979, the Central Computing Bureau in the Central Statistical Office has been providing centralized computing facilities for governmental agencies. The system is currently built around two IBM 4341 computers located at Al Jufayr, with on-line terminals at various locations in about a dozen Ministries and other governmental agencies. Following the expected recommendations of a National Committee on Computerization, there are plans to expand existing facilities to a network of three centres (with the addition of centres in the Ministry of Finance and National Economy and Gulf Technical College) and to equip each of them with an IBM 4381 computer.

At least two big national databases are being established — a Central Population Registry (CPR) that will have personal information (eventually including employment data) on every resident of Bahrain, and a National Addressing Project (NAP) that will contain exact information on where people live. Both these databases are to be updated regularly by the appropriate Ministries in the course of their day-to-day work and are intended to serve as a basic data resource for any Ministry that wishes to make use of them.

As far as physical planning is concerned, two specialized mapping systems (Intergraph) have been installed over the past year or two, one in the Survey Directorate of the Ministry of Housing and the other in the Ministry of Works, Power and Water.

The Survey Directorate system is based on a VAX 730 with two workstations, an off-line Calcomp plotter and one 675 Mb CDC disk drive. There are plans to add a third workstation and a second disk drive in the near future. The system, which is operated by two senior and two junior professionals, is being used for three main purposes:

- To maintain and produce National Survey Directorate (NSD) topographical maps of Bahrain (of which there are some 800 sheets covering virtually the entire country at a scale of 1:1000).
- To prepare a Cadastral Index Map (CIM) based on the Land Inventory Project (LIP), which is designed to consolidate information on all land surveys in the country; the Map is substantially complete now except for the cities of Manama and Muharraq where there are special difficulties.
- To create and maintain a comprehensive national cadastre of all properly surveyed plots of land; although the Directorate is

processing about 10,000 surveys per year, the cadastre is still not complete.

These activities are supported by two other computer systems currently operating in the Survey Directorate:

- 15 terminals and two printers on the central government computer system, which are used exclusively for running a custom-written programme for administering all the surveys the Directorate is asked to carry out; and
- seven Casio microcomputers (Z-80 based machines with 64 Kb RAM and two disk drives) and five plotters, which are connected to the Intergraph VAX 730 via RS-232 ports and used (with a custom BASIC programme written in-house) by the 40-odd surveyors in the Directorate for pre-processing cadastral field data and then batch-entering them into the Intergraph database.

The second planning-related Intergraph system is in the Ministry of Works, Power and Water. This system is larger than that of the Survey Directorate, being based on a VAX 785 with six workstations and one 675 Mb disk drive, but the two systems are fully compatible with each other. There are plans to add workstations and a second disk drive. Current operation requires about one senior and two junior professional staff for each workstation plus two senior system development/maintenance staff for the system as a whole. The system is used primarily to maintain records of the physical infrastructure that the Ministry is responsible for (i.e., roads, sewer lines, water mains, drainage systems, power lines and street lighting). Digitizing of records is currently about 60 per cent complete overall (100 per cent complete in respect of some plant, such as roads and water mains) and is to be finished by the end of 1986.

Moreover, the Bahrain Telecommunications Company (BATELCO) plans to install a similar Intergraph system by the end of 1986, although it will probably be based on the MicroVAX II rather than the older VAX series. Its purpose would be to help with the management of the telephone networks and facilities owned and operated by BATELCO. Naturally, this has led to the idea that all three systems could be linked together on-line in a network of some kind.

Finally, it should be noted that the Housing Bank has a Wang VS-80 computer system (which may soon be upgraded to a VS-100 system) for managing its loan records. However, these records apparently contain little data beyond what the Bank requires for financial processing and analysis of the records. There is not much physical or socio-economic information about the accounts. In addition, the Bank's computer acts as a gateway for the three terminals on the central government computer that are leased by the Ownership and Loans Directorate of the Ministry of Housing for maintaining its loan-application records.

Briefly, then, this is the context in which the Physical Planning Directorate is contemplating its information needs and resources. PPD is an actor in the national planning process in Bahrain, and other actors in the process are already taking steps to improve their information resources, including computerization in some cases; and this can be expected to continue for the foreseeable future. It is entirely appropriate, therefore, for PPD to be giving priority to the same issue at this time.

Perhaps the most fundamental point to be made is that better information for planning means more than just collecting more data. Better planning information means a systematic approach to getting data "into" the planning process. Specifically, this means:

- collecting data that are *useful*;
- making sure they are kept *up to date*;
- storing them in a *convenient and accessible form*; and
- helping planners make *effective use* of them for planning.

Moreover, these are interdependent objectives. It is unwise to concentrate, even initially, on just one of the objectives, on the assumption that the others can be taken care of in due course. Collecting data, for example, makes sense only if they can be kept up to date, they can be made readily accessible to potential users and users can be taught to make good use of the data. Information for planning, like the proverbial chain, is only as strong as its weakest link.

Thus, the main job of the Information Section is going to be development of an integrated information strategy for the PPD. On the one hand, the Information Section has a duty to take an active and even aggressive role in promoting more effective use of data for planning in PPD. The Section should know where and how to get useful and up-to-date data and how to make effective use of them — and it should be trying to encourage and assist planners in PPD to do so. Thus, its role is partly one of "promoting" new data sources and new techniques and partly one of providing support to others.

At the same time, the Information Section must be careful not to come between the four "line" Sections of PPD (Master Plans, Urban Renewal, Detail Planning and Implementation) and their information needs. The primary responsibility for identifying and collecting data has to remain with those who are going to use the data. No one should try to anticipate or, worse still, dictate to the line Sections what their data needs ought to be.

Thus, the role of the Information Section should go well beyond the passive collection of data on a "come-and-use-it-if-you-like" basis. In a sense, the basic function of the Information Section is to "sell" the other Sections on the need to devote resources to getting and using information, as well as helping them do so. Yet, in the end, the role remains essentially one of a catalyst — encouraging but not participating, demonstrating but not directing, assisting but not controlling, supporting but not supplanting the line Sections of PPD.

On this basis, the following tentative set of priorities or outline work plan for the Information Section is proposed:

1. In co-operation with each of the line Sections, make a preliminary inventory of current data needs within PPD. The purpose of the inventory would be not so much to make a list of *all* such needs as to identify some of those which are most *critical* in terms of the quality of the data and their importance to the activities of PPD. This will help the Information Section both to set its own priorities in terms of promoting improved access to information and to see clearly what methods and techniques are being used for planning in each Section.
2. In co-operation with the appropriate line Sections, select a number of areas where key data are required and work with the Sections concerned on ways of improving the collection, updating and use of relevant data.
3. On the basis of its developing experience and expertise, the Information Section should identify new methods and techniques for using information for planning and assist interested line Sections in incorporating such methods and techniques in their planning processes.

There is little doubt that information management can benefit from automation. Much of what we do with data in planning — sorting and finding, adding and subtracting, analysing and synthesizing — is boring and mechanical in nature. In principle, at least, that sort of thing can easily be done by a computer. Of course, computers cannot replace the creative skill and judgement of the planner. However, by assigning to computers some of the mundane and trivial tasks, we can free planners to devote more time and effort to tasks which really do require skill and judgement.

Thus, improved information for planning in PPD will come as much from the better use of data as from collecting more data. This means providing planners with not just more data but also more capacity to use them. If computerization is chosen as a way of doing this, it means putting computer power directly in the hands of planners themselves. If a computer specialist has to stand perpetually between the computer and its users, then almost inevitably the uses become standardized and inflexible. Improved information use will come not just from providing large-scale databases but also from enhancing the personal information-processing capacities of individual planners in PPD.

Thus the important factor here is not the power or capacity of the computer as much as its accessibility to potential users. There are at least four important dimensions to accessibility:

- the physical number and location of computer terminals;
- the ready availability of off-the-shelf software;
- the ease of use ("user-friendliness") of the system; and
- the availability of training and other forms of support.

Given these objectives (to say nothing of budgetary limitations), the most appropriate form of computer hardware for PPD at this point in time is almost certainly the microcomputer. Microcomputers are easy enough for non-specialists to use and numerous enough for planners throughout PPD to have the opportunity to do so. For most purposes, standard IBM-compatible microcomputers with dot-matrix printers will be adequate. (If desired, bilingual capability for the computer and the printer can also be specified.) As far as numbers are concerned, it will be hard to ensure even minimal computer access throughout PPD without at least two workstations per Section, i.e., ten in all.

As far as peripherals are concerned, PPD may want in a second stage to add plotters and digitizers to its microcomputers, although mapping on a basic microcomputer is still relatively limited in comparison with minicomputer-based systems such as the Intergraph. Similarly, PPD may want to add hard disks to some of its microcomputers in due course; in general, the best way to do this is by means of external disks with built-in backup units. Local area networking is probably unnecessary for the present, although provision should probably be made for communication with the Intergraph systems in the Survey Directorate and the Ministry of Works, Power and Water as well as with the Central Computing Bureau mainframe network. Off-the-shelf software for word processing, spreadsheet analysis, database management, project management and BASIC programming should be sufficient to meet most planning needs for at least the first few years — especially if it is supplemented by a good library of public-domain, planning-related software.

Training and support will clearly place a considerable burden on the Information Section, especially in the early days; provision should be made for strengthening its resources in this respect. Four specific recommendations are made here:

- If possible, training should be provided in the context of job-related needs; the aim is to help planners become the best possible planners, not computer specialists.
- Secondly, in job settings, it has been found that (given the opportunity to do so) people will often do a lot to help themselves and one another to learn how to use microcomputers. So the important thing about training is to lay a good foundation and then provide the opportunity and the motivation to encourage people to learn on their own.
- Thirdly, PPD will probably need at least one full-time microcomputer specialist to assist staff with the development of specific applications of generic programmes such as database management systems, spreadsheet analyses and simulation models, and to assist with downloading data files from foreign computer systems.
- Finally, it is important also for senior staff to increase their awareness and understanding of the potential for information use in PPD, particularly in the light of the "revolution" that has occurred in the past five years or so, owing to the development and widespread use of microcomputers.

Case 14

Setting priorities (Turkey)

The Turkish Ministry of Public Works and Settlement was formed in 1983 by merger of the Ministry of Public Works and the Ministry of Reconstruction and Resettlement. It is made up of four "operating units", two "annexed units" and two "ancillary units". The following summary (taken from a 1986 Ministry publication) gives an idea of the range and diversity of its activities:

- The General Directorate of Construction Affairs, which has seven Departments, about 28,000 employees (half of them located in Regional Offices) and a capital budget of $450 million in the current fiscal year.
- The General Directorate of Railroads, Harbours and Airports Construction, which has 15 Departments and is responsible for investments worth more than $2 billion.
- The General Directorate of Disaster Relief, which has six Departments, a capital budget of $4.5 million in the current fiscal year and administers programmes and projects worth a total of $73 million.
- The General Directorate of Technical Research and Implementation, which has eight Departments, 625 employees and a capital budget of $6.4 million. This Directorate allocates about $60 million in grants to municipalities, licenses some 25,000 civil contractors and allocates and disposes of approximately 15,000 housing units annually.

The "annexed" and "ancillary" units in the Ministry are:

- The General Directorate of Highways, which has 12 Departments, 17 Regional Directorates and 40,000 employees; and graded, asphalted, macadamized or repaired more than 10,000 kilometres of roadways and built 112 bridges and 1,118 other structures during 1985.

- The General Directorate of the Land Office, which has six Departments, 204 employees and a capital budget of $18 million; and manages approximately 8,445 ha of publicly owned land, including 28 housing developments and 66 commercial and industrial estates.
- The General Directorate of the Bank of Provinces, which has 13 Departments and a capital budget of $170 million and (in 1985) financed 1,152 municipal and rural development projects and made $27 million worth of short-term loans to 500 municipalities.
- The General Directorate of the Real Estate Credit Bank of Turkey, which has 6,000 employees in 300 domestic branches, holds more than $500 million in deposits and provides more than $400 million of financing for more than 400,000 housing units.

Thus the total staff of the Ministry (including the Regional Offices) is about 80,000.

It is clear that such a large and active Ministry provides many extensive opportunities for realizing the benefits of computerization. Indeed, it takes little imagination to think of potential applications in all parts of the Ministry for computer facilities, such as word processing, database management, project management, accounting, statistics, engineering analysis, architectural design, mapping, seismic risk analysis and economic and physical modelling. Of course, some computer applications in the Ministry do already take place and, in some cases, have done for years; but, with developments in technology over the past five to ten years and the consequent fall in the price of computer hardware, the demand for such facilities in the Ministry could well explode in the next few years (*Development Plan* 1985, pp. 177-8).

To meet these needs, a Data Processing Department and a Computer Centre were set up in 1985 as a Department within the General Directorate of Technical Research and Implementation. The Department is responsible for providing data-processing services to units throughout the Ministry and for collecting data in respect of services rendered by all of the units within the Ministry. The Computer Centre was equipped with three microcomputers (an IBM PC with 64 Kb RAM and 1 disk drive; an IBM PC with 256 Kb RAM and 2 disk drives; and an IBM PC-XT with 256 Kb RAM, 1 disk drive and a 10 Mb hard disk), three printers and software for word processing, spreadsheet analysis, database management and project management. The Department has subsequently obtained copies of other software but lacks manuals and other training aids.

At the end of 1985, the Department also took delivery of a Data General MV-4000 Eclipse minicomputer. The system is configured with 2 Mb RAM, two disk drives with a capacity of 350 Mb each, a magnetic tape backup unit and a 16-port asynchronous input-output controller. This last is currently being used to connect 15 monochrome terminals (without graphics) and one printer, all of them situated in the Computer Centre. The capacity of the system can be increased to up to 64 ports, 8 Mb of RAM and 9.4 Gb of on-line disk storage. The Centre also has a high-speed line printer (1200 lpm) and a teletype terminal but they are not currently connected to the system. The MV-4000 uses AOS/VS (a multi-user, multi-tasking time-sharing operating system) and has FORTRAN, COBOL and BASIC as well as INFOS, SQL, Trendview and various other database management utilities.

In terms of applications, the Department's first priority has been database management. Four applications have been identified so far:

> *Municipal grants monitoring* This database contains records on more than 10,000 disbursements under some 5,000 different projects in about 1,700 municipalities. The database was originally designed and programmes first written on microcomputer by departmental staff using dBaseII, but performance was too slow. Without access to dBase enhancements (a compiler or dBaseIII or III+) or alternative microcomputer software, departmental staff have converted the programmes to COBOL (and the database to ISAM files) and transferred them to the MV-4000, where they now run under INFOS.

> *Roster of civil contractors* This database consists of data on all the approximately 29,000 licensed civil contractors in Turkey. Programming (COBOL) is being done under contract by the supplier of the minicomputer system. The database is projected to reach 500 Mb in size but only about 2.5 Mb (dealing with some 1,782 contractors) has been keyed in so far; so the database is not yet complete or operational.

> *Personnel management and payroll* The Department is to prepare a comprehensive employee database for the Ministry to facilitate personnel management and payroll. In the meantime, the Department has completed a prototype payroll system for the staff of its own General Directorate (with approximately 625 people).

> *Construction materials* The Department is to establish a database on construction and building materials used in Turkey. The preliminary systems analysis for this project has been completed, but the programming has not started.

The next priority for the Department is graphics applications, possibly including both mapping and computer-aided design (CAD). However, specific programming targets and their users have not yet been firmly identified. Beyond that, the Department is also contemplating computer applications for land management, urban research, simulation models and engineering analysis. For 1986, the Department has decided to concentrate its capital spending on development of the MV-4000 system, with priority being given to the following purchases:

- Expansion of system RAM to 4 Mb.
- Addition of a third disk drive (592 Mb) and cabinet.
- Addition of 10 terminals (five to be connected remotely by modem) and a second 16-port controller.
- Addition of two medium-resolution colour/graphics terminals.
- Purchase of one A0-size and two A3-size plotters (one of the latter to be of the inkjet type) and a digitizer.
- Purchase of appropriate software for graphics, statistics and engineering applications.

The only planned expenditure on the microcomputer network is expansion of RAM in all three microcomputers to 512 Kb, purchase of one colour/graphics monitor and supply of appropriate software for graphics.

Based on the foregoing summary, there are at least two critically important features of the computerization strategy currently being followed in the Ministry of Public Works and Settlement.

First, there is the decision to co-ordinate the process of computerization throughout the Ministry from the position of a unit located within one of the General Directorates.

Secondly, there is the decision to acquire and support both stand-alone microcomputers and a multi-user, time-sharing minicomputer system.

Both of these decisions will present a considerable challenge to the skills, both organizational and technical, of the staff of the Data Processing Department. The success of the Department in fulfilling its mandate will depend to no small degree on its ability (a) to work outside its own General Directorate as well as within it and (b) to maintain an effective balance between the "micro" and "mini" systems.

Co-ordination within the Ministry

As noted previously, the Data Processing Department could face an explosive growth in demand for computer services in the next few years. This demand could far outstrip the resources and capacities of the Department to meet it. In fact, in a Ministry of 80,000 employees, the demand for word processing alone, aside from any other uses, could swamp the capacities of the Computer Centre — in terms of the number of its terminals, the capacities of its memory and processor, the size of its disk storage and the ability of its staff to provide training and support. In other words, if the Data Processing Department tries simply to meet all of the Ministry's computer requirements through its own resources, those resources will certainly prove inadequate. If, on the other hand, the Department can develop a variety of relationships with computer users and potential users that demand different degrees of commitment and support, then it stands a good chance of playing an effective role in computerization on a Ministry-wide basis.

One element in the success of the Department on a Ministry-wide basis will be its ability to develop an outward-looking approach. This means (so to speak) not just sitting back and saying, "Well, here we are, ready to help anyone who comes along and asks us to do some computing!", but taking the initiative of going out to see what opportunities there are for computerization and how they can best be exploited. Some people will no doubt come to the Department asking for help, but a lot of others will not. Some people will not come because they do not imagine that computers could be of much use in their work. Others will not come because they do not want to become dependent on another Department or perhaps because they think the Department will "take over" their project or fail to provide them with the service they really want. Such feelings may be unjustified, but they may nonetheless exist.

It is vital, therefore, for the Data Processing Department to go out into the Ministry saying that computers now offer opportunities for almost everyone and that the Data Processing Department stands ready to assist in a variety of ways according to needs. This could range from a full on-line link to the central computer, through various forms of periodic communication with and assistance from the central computer, to no more than advice and support in establishing an independent (but compatible) facility. An active and "non-threatening" approach of this kind will allow the Department to retain an influence over the "shape" and pace of computerization in the Ministry that far exceeds what it can actually do by itself with its own resources. Certainly, it will be futile to try to compel people to use the facilities of the Data Processing Department when they do not want to. The experience of organizations where traditional "DP" Departments have tried to impose their services bears ample testimony to its failure, especially now that improvements in technology offer the simple, low-cost alternative of the micro-computer. DP Departments that do not meet their users' needs soon lose their "clients".

Meeting user needs is a complicated business that goes far beyond the technical specifications of a programme or a file structure. It entails a familiarity with the user's professional field as well as a certain psychological appreciation of the user's style and approach. No one knows these better than the user. So the best approach to meeting user needs is to educate and involve the user as much as possible in the process of design and implementation of a system. In fact, the more the user can be encouraged to do for himself or herself, the more the user will learn about computers and the more effectively the user will be able to take advantage of the facilities provided.

In short, the key for the Data Processing Department establishing an effective kind of co-ordination over the process of computerization in the Ministry is development of a strategy that is "pro-active", not just "re-active". That is, the Department should go looking for work and not just wait for work to come looking for it! The Department will succeed in influencing computerization to the extent that it is willing to go out into the rest of the Ministry and "sell" both the idea of computing and the flexibility of the services and support that it can provide. While offering advice and help, moreover, the Department can encourage other units in the Ministry to get their own equipment and to train their own manpower. Thus, the Department will minimize the commitment of its own equipment and manpower to any one project while extending its support for and influence over computerization throughout the Ministry.

In this way, the Department can help to co-ordinate hardware, software and training on a Ministry-wide basis. It can promote standards of compatibility for communication within the Ministry. It can encourage data exchange and software sharing. It can help establish standards for operation and training. It can act as a focal point for contacts with users in other Ministries. However, its effectiveness in these roles will depend on its ability to take and keep the initiative; this kind of leadership cannot be assumed or even mandated.

A balance between micros and minis

There is a tendency to think of computers as all more or less the same; so, if something has to be computerized, you can do it with whatever spare computer capacity there is available. This, however, is a mistake. Different computers have different capabilities, strengths and weaknesses; and these make it better or worse in certain kinds of circumstances. For example,

- Microcomputers are relatively easy for non-specialists to learn how to use, thanks to the variety and sophistication of the available software.
- Microcomputers are powerful enough to be able to meet many of the Ministry's requirements, from word processing to design to engineering analysis, at least for an interim period until users learn

to go beyond their capabilities. There is a lot of off-the-shelf microcomputer software available to help meet the Ministry's needs.

- Microcomputers are flexible and portable, so they can easily be moved from one office to another as required.
- Microcomputers provide redundancy in the event of equipment failure. If one microcomputer fails, it does not affect any of the others.
- Microcomputers are cheap enough for access to be provided on a widespread basis at a reasonable cost.
- Microcomputers do not need expensive support facilities; in addition, parts and service are not as expensive as they are for large computers.
- Microcomputers can share data and software just by physically exchanging disks. Electronic communication by modem or local area network (LAN) is also possible at relatively low cost.

On the other hand, for large databases, for high-speed or heavy-duty number-crunching, and for applications where several users want simultaneous on-line access to the same database, minicomputers and even mainframes are appropriate. In general, microcomputers represent a sensible first step towards computerization, but there are some applications that will exceed their capacities. Moreover, microcomputers can help relieve pressure on large machines by freeing them from more trivial applications. Experience with microcomputers also means that users will graduate to sophisticated computers a lot more informed than they otherwise would be. Finally, microcomputers will help create a large user base of computer-literate staff in the Ministry who will continue to push for and support increased computerization.

So much for the supply side of the equation. Assessing the demand side is not easy, especially when users have relatively little conception of what computers can and cannot easily do. Take the case of a 500 Mb database, for example: certainly, a database that size will be easier to manage by computer than by hand, but are the benefits of computerization really going to be commensurate with its costs? Or would it be wise to rethink the purpose of the database and how it is managed?

Consider, for example, the fact that just entering 500 Mb of data into a computer is probably going to take at least 50 person-years with access to 50 terminal-years. This does not include the time required to extract the data from files and other documents where they may be and to present the data in a form in which they can be entered at the rate just implied (10 Mb per person per year) — a process which will itself probably take longer than the actual data-entry. Of course, these costs are highly "visible" when a database is just being set up. They become less obvious when data-entry occurs gradually, when it is integrated with normal office routine, when it is done at or close to the point of original collection of the data, when it is decentralized to regional or provincial offices and so on. Nonetheless, it is tempting to re-examine the whole undertaking when such costs are involved and to consider whether, in the light of certain precise data-processing requirements, most of the desired benefits of computerization could be achieved at much reduced cost through a hybrid system in which existing paper records were supported by limited computerized data processing. In short, how much of the 500 Mb of data will actually be processed (e.g., printed on official documents, used as a search criterion, processed mathematically, analysed factorially, regressed linearly or cross-tabulated) and how much of the data is there just "because someone might be interested in it"?

Because of the complexity of assessing the demand for and proposing the supply of computer facilities, it is essential that the Data Processing Department continue to strengthen its capacities.

First, the Data Processing Department should be clearly mandated by the Ministry to establish a mechanism for identifying, reviewing and co-ordinating computer needs throughout the Ministry. This mechanism would be concerned not just with hardware and software but with training and systems-analysis needs as well. The same mechanism could also recommend standards and priorities for developments in all these areas and promote common services and facilities, such as communication networks, computerization of regional offices and links with governmental agencies outside the Ministry as well as some of the larger municipal offices.

Secondly, as emphasised above, it is essential to maintain a balance between the "micro" and "mini" systems if each is to continue to complement the other. Given the pattern of capital expenditure to date and projected for the balance of 1986, it is clear that the microcomputer facilities are going to need strengthening. It is particularly important that these be deployed in such a way as to maximize access for the largest number of people and to the widest variety of software relevant to the activities of the Ministry. In addition, priority should be given to establishing a link between the two systems, so that users and uses can be shifted from one system to another as their respective needs and capacities change.

Two of the most important areas where the Data Processing Department can provide advice are those of training and systems analysis. In both these areas, evaluation of the Ministry's needs requires skill and judgement. Moreover, in both these areas, the Ministry is dependent on outside advice from people and companies who stand to benefit from the nature of that advice (i.e., suppliers of computer equipment and services). It would probably be wise for the Ministry to have the capacity to make its own determination of needs in these areas.

Case 15

Increasing productivity (Swaziland)

Primary responsibility for human settlements planning and management in Swaziland lies with the Ministry of Natural Resources, Land Utilization and Energy, which was formed in 1983 as part of a general re-organization of government. Under the *Fourth National Development Plan, 1983/4–1987/8*, the Ministry is assigned a broad range of tasks related to the development of human settlements throughout the country. These tasks include:

Providing land-capability assessments and plans for land and water development projects;

Planning for the most economical and efficient development of urban and rural settlements throughout the country;

Providing land-surveying and mapping services in all parts of the country; and

Undertaking programmes and projects for improved housing and housing infrastructure for low-income and middle-income groups.

Significantly, the Ministry is expected to achieve all these objectives without any increase in its recurrent budget during the five-year period of the Plan. Capital expenditures are actually expected to decline over the same period. Thus, it is clear that any improvement in performance by the Ministry is going to have to come not from increases in staff but from improvements in productivity.

The Ministry is divided into a number of sections and branches, each with specific powers and responsibilities. Six of these units are actively involved in human settlements planning and management, namely, the Physical Planning Branch, the Land Use Planning Section, the Housing Branch, the Survey Branch, the Deeds Registry and the Lands Branch. The functions and resources of each are discussed below.

Physical Planning Branch

According to the current National Development Plan, the Physical Planning Branch is responsible for "the planning and guidance of development over [*sic*] the physical environment . . . at national, regional and local levels, by making the fullest, most appropriate and efficient use of available human and natural resources".

The Plan assigns to the Physical Planning Branch a comprehensive set of priorities for the next five years:

- to prepare a National Physical Development Plan for the country;
- to update all physical planning legislation;
- to prepare Regional Development Plans;
- to formulate environmental protection codes;
- to prepare urban structure plans;
- to prepare housing designs and site layouts; and
- to decentralize its own physical planning functions.

These priorities are to be "the future source of all new project proposals" and thus are to anticipate "what has to be done and the order of sequence to be adopted to accomplish the . . . stated objectives".

The Physical Planning Branch has a broad mandate to plan, control and co-ordinate the development of the physical environment at national, regional and local levels. The legal foundation for this mandate lies in the Town Planning Act of 1961, which calls for preparation of "Town Planning Schemes" for all cities and towns (of which there are now ten). Although plans have been prepared for most of these, none has yet been approved. In addition, the Building and Housing Act extends the authority of the Branch to all public improvements to the built environment which do not come under a Town Planning Scheme. The 1961 Act is to be amended and largely superseded by a new Human Settlements Act, which will extend planning to all forms of human settlement above a certain size and will significantly increase the planning authority of the Branch.

The Branch has a very small staff (including only three planners) and depends on private consultants for most of its professional work. For example, the National Physical Development Plan is currently being prepared for the Branch by a consulting firm based in Nairobi, Kenya. Even the regular work of the Branch, including review of building-permit applications, is growing more quickly than the capacities of the Branch to deal with it. With urban population growth estimated at 10 per cent per annum and a housing shortage that already affects about 30 per cent of the low-income population, there is little respite likely in the need for immediate-term planning. At the same time, the Branch is aware of the need to take an active role in long-term planning and to reduce its dependency on private consultants. The only possible way to meet these conflicting demands with existing resources is through increased productivity.

Land Use Planning Section

The Land Use Planning Section (until 1984 part of the Ministry of Agriculture) has responsibilities similar to those of the Physical Planning Branch, but its jurisdiction is rural. The principal duties of the Section include the selection and designation of Rural Development Areas — rural areas with good agricultural potential, a growing population and, thus, increasing development pressures — and the preparation of plans for rural development projects. Thus, the Section emphasises natural resource-based planning, using information on soils, vegetation and water resources as well as on rural population.

The Section has a staff of eight professionals and eight draughtsmen and technical assistants. There is one Apple II microcomputer (with printer) in use. It is used mainly for water-resources planning, especially for the storage and analysis of hydrological data in rural areas.

Housing Branch

The number of people in Mbabane without adequate shelter may already be as high as 30 per cent of the total population. Most of these people are living in "informal" housing in squatter settlements on the fringe of urban areas. As rural dwellers continue to see advantages to living in urban centres, the population shift to the cities will continue, and the urban housing problem will remain acute. The situation in Mbabane, which is surrounded by steep hills, is particularly difficult.

National responsibility for the delivery of housing is divided between the Ministry's Housing Branch and one of its parastatals. The Housing Branch looks after housing for low-income groups by implementing programmes to provide basic infrastructure and assisted self-help housing. Public housing for middle-income groups consists primarily of rental flats which are built and operated by the Industrial Housing Corporation (IHC). Full capital-cost recovery and even some surplus is expected from IHC projects.

As an indication of the high priority accorded housing under the current National Development Plan, the Housing Branch and the Industrial Housing Corporation are to be combined in a National Housing Board. The new Board will be the sole agency responsible for all public housing throughout Swaziland. Here again, meeting new responsibilities with existing staff can be achieved only through improvements in productivity.

Survey Branch

The Survey Branch is responsible for all governmental survey work, mapping (including continuous revision of the basic national 1:50,000 series of 31 sheets) and maintenance of the national cadastre. The Branch is also responsible for approving all survey work done by the private sector, in order to ensure the accuracy of the national cadastre which is (in turn) the foundation for the land tenure system. All official cadastral survey records are held by the Branch, which provides copies to other governmental offices and individuals on request.

The Survey Branch has one Epson QX-10 microcomputer (eight-bit) and an Epson dot-matrix printer. The system has been used for several years to do survey calculations, such as checking the co-ordinate geometry of survey data from the field. Software has been developed by a local private surveyor who had developed and tested the system for use in his own practice. There is no computer staff as such in the Branch, but there are now several people who know how to use the equipment.

Deeds Registry

The Deeds Registry provides for the registration of all title deeds and transfers of ownership pertaining to land, including ordinary transfers, transfers by estate, public auctions and registration of mortgages. Certain other private contracts are also registered here, as are Court Interdicts and all transfers of land from the Crown. The Registry also carries out searches on title and can advise whether particular individuals (e.g., applicants for governmental housing lots) are property-owners.

The Registry has a total staff of fifteen. Their annual workload consists of some 500 transfers of ownership, 500 registrations of mortgage, 100 transfers of Crown Land and various other actions.

Lands Department

The Lands Department is responsible for the purchase and sale of all publicly owned land. In particular, the Department co-ordinates the supply of serviced land for housing through the allocation and sale of plots to landless individuals who can show need. Costs are partly recovered through a repayment scheme administered by local authorities. The delivery of serviced housing lots follows a precise process:

The Physical Planning Branch plans the site, including lot and street layout as well as water, sewer and electrical infrastructure.

The Survey Branch surveys the property, lays out the plots, marks the boundaries and updates the national cadastre.

The Land Valuation Section assesses the new plots.

The Lands Department screens applications received by local authorities to exclude persons who already own property (as shown in records at the Deeds Registry) and people who cannot show real need.

The Allocation Committee reviews the approved applications and allocates the available plots.

Local authorities assume responsibility for administration of urban services and collection of repayments related to all plots in the site.

The successful applicant secures his or her plot by making an initial payment to the appropriate local authority of one-quarter of the total costs (land value and surveying costs) and agreeing to a subsequent repayment schedule.

The Ministry formally registers the transfer of ownership in the Deeds Registry.

Administration of a complex, interagency process such as this is quite difficult. As the number of lots required for housing inevitably rises, pressure on the resources of Lands Department staff will also increase.

In short, the Ministry has the authority and the resources necessary for planning and managing the physical environment, including its human settlements. Although the Ministry is relatively young, and the emphasis on human settlements relatively new, the Ministry is playing an increasingly important role in policy formation, project execution and development control. However, growing duties mean a growing burden for Ministry staff, which is therefore looking for ways to improve productivity. The use of microcomputers for certain technical tasks in the Land Use Planning Section and the Survey Branch has already demonstrated how much microcomputers can help. The time is ripe to expand their application to a broad range of responsibilities within the Ministry. Given the increasing workload in the various branches of the Ministry and the "freeze" on additional staff for the foreseeable future, extensive use of microcomputers can be justified on the grounds that they can help improve productivity in both administration and planning.

It should also be noted that the two existing microcomputer systems in the Ministry are vulnerable to hardware failure. There is only one microcomputer in each system and each is incompatible with the other.

Thus, if one were to break down for any reason, its work would be interrupted and all its data would be inaccessible until repairs could be effected — which might be a matter of weeks or even months. Quite apart from expanding their use, therefore, the Ministry would be wise to provide backup capability for existing uses and to budget for repairs and maintenance.

Additional uses for standard microcomputer systems within the Ministry could include any of the following:

(a) *Monitoring the status of applications* for development permits or allocation of housing plots, as these applications proceed through their various stages of review or approval. For example, as requests are received, each is entered into a database that acts as an "electronic logbook". As each application moves through the various stages of review or approval, its record is updated. When approval is granted or denied, the record is transferred to an archival database to preserve the historical and statistical details of each application. The active and archival databases could be queried for information and statistical data. For example:

How many applications were received during the first quarter of 1986? How many were approved? What is the status of a particular application? How many non-residential applications are pending?

Some features of the database could be automatically monitored, and the system could be designed to generate letters or memoranda from time to time (e.g., every two weeks) to inquire about or advise on the status of outstanding applications.

The resulting database would be a useful source of information on development trends. For example, given the right design and structure, the database could show the distribution of applications by proposed use, land area, value, etc.; or even the time taken at various stages in the approval process, either for any particular application or for a selected group of applications. This would be useful for administrative purposes to identify bottlenecks in the approval process.

(b) *Performing financial and demographic modelling.* With standard spreadsheet software, microcomputers can be used to help prepare project and departmental budgets, to create capital-cost models for construction projects such as sewer and road systems, and to make population projections.

(c) *Running housing models* such as the Bertaud Urban Cost Model and the US AID Housing Needs Assessment Model. The Bertaud Model was created to optimize the design of affordable low-cost housing schemes for low-income populations. The model facilitates analysis of trade-offs between various land-use and cost parameters and testing of the affordability of different mixes of lot sizes and types, standards of infrastructure and schemes for cost-recovery. The US AID Housing Needs Assessment Model is designed to forecast national housing needs, given the quantity and quality of existing housing stock and the future demand for housing due to natural population increase and migration. The model estimates the total investment required to achieve specified minimum standards of affordable housing.

(d) *Physical planning and analysis*, including land-suitability assessment and optimum-location analysis — both of which are helpful in deciding where to build central facilities such as schools, clinics, markets and retail centres. There are several programmes available which can find optimal locations on the basis of data about the distribution of population, the nature of existing transportation networks and the location of existing facilities. Among such programmes are two available from UNCHS (Habitat), Urban Data Management Software (UDMS) and GRID.Micro.

(e) *Statistical analysis* of basic socio-economic data such as personal income, housing stock and household expenditure. Microcomputer-based statistical packages are now available to enable the Ministry to analyse not only standard census data but also data from special surveys and questionnaires.

(f) *Business graphics*, such as bar-charts and line graphs. Graphics can enhance the presentation of statistical and financial data used in reports, proposals and project budgets. Popular programmes can operate directly on data produced from many of the above applications, such as spreadsheets, databases and statistics packages.

(g) *Word processing*, especially in connection with official Town and Regional Plans. Word processing is useful not only because of its speed but also in cases when the content or structure of a document is likely to be changed frequently or radically during its creation. With a word processor, revised drafts can be produced without laborious retyping of unaltered material. This also prevents the introduction of new typing errors into old parts of a draft. Similarly, for standardized legal documents, such as planning notices or building permits, word processing makes it possible to type in the details of a specific case (owner's name, location, lot number, etc.) and then to "merge" that information with the legal text required for such cases that is already stored

> on a disk. Indeed a whole library of standard paragraphs (legal definitions, land-use descriptions, planning terms, etc.) can be stored on disk and used over and over again without ever having to retype them.

In these and other ways, it is clear that microcomputers can make a significant contribution to the operations of the Ministry. Preliminary experience has been entirely positive and, in view of restrictions on future staff recruitment, prospects are equally good for continued improvements in productivity through computerization.

On the other hand, computerization is not a panacea. There are even dangers in proceeding too far too fast. Systems are not flexible enough to adapt to changing needs. People do not get enough training to make them efficient. Organizations do not get enough time to assimilate the new technology and learn where and how best to use it. In the words of a 1982 study for the Ministry,

> big projects . . . (such as a comprehensive planning information system for the Ministry) do not appear appropriate . . . because the physical planning process is too ill-defined. . . . [Conditions] do not argue for a massive attack on the data management problem; rather, one should advance in several stages.
>
> (UNCHS, 1982, pp. 15 and 27)

Thus, it is important to start with a practical set of achievable goals — productivity tools that Ministry staff can make immediate use of. Then when staff have gained a working familiarity with basic hardware and software, sophisticated applications can be proposed and (if necessary) additional equipment provided. This approach has two advantages: (i) immediate results in terms of improved productivity and (ii) long-term benefits in terms of the motivation and experience gained by Ministry staff. Both provide an invaluable foundation for implementing sophisticated computer applications in the future.

Note

This case was prepared with the kind assistance of Mark Richard Brown, UNCHS (Habitat) staff member, who undertook the mission on which the case is based.

Case 16

Improving computer use (Jordan)

The Housing Corporation of Jordan was established in 1966. Its responsibility is to provide housing for low-income and middle-income Jordanians in accordance with national objectives, as set out in the current multi-year plan for economic and social development. To date, the Corporation has delivered some 17,000 units of housing in more than 70 different projects all over Jordan. This represents about 1.5 million square metres of residential floorspace at a total cost (unadjusted) of about Jordanian Dinars 100 million. This is estimated to be between 10 and 20 per cent of the total housing production in Jordan over the past 20 years. Detailed figures for the period 1969 to 1983 are presented in Table 13.

Over this period, the total budget of the Corporation is estimated to have been about Dinars 170 million. Almost all of this amount was used for capital expenditure; less than 2 per cent went on recurrent (or operating) expenses. To date, the Corporation has been financed from three main sources: direct government contributions (20%), domestic loan capital (45%) and foreign direct assistance (20%). About 10 per cent of the Corporation's investment has so far been repaid by its beneficiaries, but this ratio will naturally increase as time goes by. According to a study done in 1981 (Al Fandi 1981), the Corporation could be financially self-sufficient — in the sense that revenue from repayments and other sources would be sufficient to finance construction of new units without recourse to subsidies or loans — by as early as 1988. In the light of revenue estimates for 1982 and 1983, however, this now seems optimistic.

The Corporation is on the verge of completing its largest project ever — Abu Nuseir New Town — which was begun in 1978 and will create some 3,650 units of new housing. Moreover, Corporation projects have hitherto been directed at specific groups of employees (e.g., teachers, airport workers or workers in a particular industrial estate). Abu Nuseir may then represent not only a quantitative development in the role of the Housing Corporation but a qualitative one as well.

Table 13 Housing stock delivered by the Housing Corporation of Jordan, 1969–83

Year	Projects Completed	No. of Units Delivered	Total Floor Space	Mean Unit Floor Space	Total Cost All Units	Cost per Unit	Cost per Square Metre
			(Sq Metres)		(Current Jordanian Dinars)		
1969	2	286	40,354	141	902,690	3,156	22.37
1970	1	180	8,043	45	182,050	1,011	22.63
1971	2	181	11,857	66	258,023	1,426	21.76
1972	8	526	39,503	75	915,000	1,740	23.16
1973	8	301	22,965	76	738,607	2,454	32.16
1974	9	605	38,767	64	1,080,000	1,785	27.86
1975	6	370	41,005	111	1,492,000	4,032	36.39
1976	3	2,980	213,880	72	11,560,580	3,879	54.05
1977	2	89	10,819	122	753,323	8,464	69.63
1978	4	118	11,232	95	750,427	6,360	66.81
1979	6	1,192	98,172	82	5,332,895	4,474	54.32
1980	5	1,713	195,226	114	13,266,143	7,744	67.95
1981	5	4,627	549,316	119	52,129,000	11,266	94.90
1982	4	656	69,362	106	5,737,000	8,745	82.71
1983	5	459	33,514	73	3,618,000	7,882	107.95
Total	70	14,283	1,384,015		98,715,738		
Average per year		952	92,268	97	6,581,049	6,911	71.33

Source: Housing Corporation, The Role of the Housing Corporation in the Housing Sector (Amman: mimeo, 1981), tables 14 to 15, corrected for errors in translation and adjusted as noted above. Data for the past three years are taken from the draft 1983 annual report of the Housing Corporation. Data for 1984 and 1985 are not available.

Note: Data do not include the cost of loans or plots provided without housing in 1969 and 1970 (176 units at a cost of JD257,897 and 41 units at a cost of JD125,392 respectively); nor is any allowance made for housing under construction but not yet delivered or for the cost of schools or other buildings.

A few years ago, on advice from governmental computer specialists, the Housing Corporation decided to acquire a computer system and to computerize two of its most important functions, namely, management of all its outstanding loans to beneficiaries and maintenance of a general ledger for the entire Corporation. In 1983 tenders were called, and three bids received. By September, the successful bid had been chosen and a contract signed. Delivery was completed by March 1984 and installation and training shortly afterwards.

The contract provided for the supply and installation of hardware and software. The hardware consists of the following eight pieces of equipment plus associated cabling and connectors:

- one PDP-11/23-PLUS computer (Digital Equipment Corporation) with true 16-bit architecture, RAM memory of 0.5 Mb (expandable to 4 Mb) and six (expandable to 64) RS-232 serial ports;
- four 25-line, 18x20 dot-resolution VT-52 terminals (Zentec ZMS-90), with a fully bilingual Arabic-English character set in ROM;
- two CDC 9620 hard-disk storage drives (each with 80 Mb [67 Mb formatted] capacity), a DC02A disk controller capable of supporting up to four drives, and four disk-packs; and
- one high-speed, dot-matrix printer rated at 300 lines per minute (Printronix LP11-300), with a partly bilingual (lower-case English is not available) character set.

The system is supplied with the RT-11 operating system and the TSX time-sharing extension. This allows for multi-user/multi-tasking operations and is supplemented by a set of utilities for text editing, forms management and file sorting and merging. The following high-level programming languages are also available on the system: FORTRAN, COBOL, DIBOL (a DEC version of COBOL), BASIC (interpreter only) and MACRO (for assembly-language programming). Finally, the computer system includes two menu-driven programmes custom written in DIBOL for the Corporation.

The first of these is called the Loans Repayment Programme and it maintains a record of the financial status of every housing unit on the Corporation's books. On occupancy, each beneficiary makes a down payment of at least 10 per cent of the total cost of the unit (including land and services). For the balance, a fixed monthly charge is calculated based on the amount required to amortize the loan over a given term (not more than 20 years) at a rate of 5.5 per cent per annum. However, when it is received, each monthly payment is applied in full to reducing the amount of the principal. Simple interest on the outstanding balance is calculated (also at 5.5 per cent per annum) and added to a separate account; this account does not become due (and subject to interest in its turn) until the principal is fully paid off. Lump-sum or advance payments may be made at any time without penalty, although this does not normally lead to recalculation of the monthly payment. In many cases, payment is made through payroll deduction (which the Corporation in fact has the power to require from beneficiaries who are governmental employees) but payment may also be made at any branch of the Housing Bank.

The purpose of the Loans Repayment Programme is to replace the manual system that had been used for keeping track of loans and payments on all of the units delivered by the Housing Corporation. The programme currently manages data on about 17,000 housing units and is estimated to process about 7,000 transactions a month. The programme records payments, calculates interest and updates beneficiary records. There are also several other routines for sorting, indexing, analysing and printing individual records or lists of records. It is estimated that about 1,000 loans have been fully paid off. Some of these may not have run the full term (e.g., in cases where beneficiaries become entitled to interest-free loans from their employers which they can use to pay off the Housing Corporation loan).

The second custom programme is a General Accounting Programme which is designed to record all payments to and by the Corporation, including details of the date, account, nature and amount of the transaction. The list of accounts currently contains about 250 codes. The purpose of the programme is to provide the Corporation with up-to-date information about the financial state of its affairs at all times. The programme can provide trial balances, monthly balances and transaction summaries by time period or by account number. As such, the computer programme does not really replace a manual procedure (as the Loans Repayment Programme did). The manual recording and processing of payments still goes on. Instead, the General Accounting Programme duplicates part of the manual procedure in order to provide quick access to some of the information it contains.

The computer system is situated in the basement of the headquarters building in Amman and is operated by the Accounting Section in the Treasury Division. The system has been running for about a year now (Hobbs metres on the disk drives show just over 1,700 hours of use to date) and is reported to have been largely trouble-free. No records are kept of users or uses. For the first three months or so, both the Loans Repayment Programme and the existing manual procedure were in operation, until the Division was satisfied that both were producing identical results. Now, the manual records are no longer maintained and the Corporation is wholly dependent on the computer. The General

Accounting Programme is also in operation, but (as already noted) the manual system remains in full operation.

Present uses

Two basic issues need to be addressed in connection with present uses of the computer system: how well were the Corporation's computer needs identified in the first place (systems analysis) and how well are those needs being served by the present hardware and software (systems design)?

Turning first to the question of identifying needs in respect of loans management and general accounting, it is notable that a generally narrow interpretation was put on both these activities. Now it is always possible to argue for the expansion of computer activities in the name of research or planning; but, even if we concentrate on the loans and cash transactions respectively, the conception of what it means to manage these activities seems surprisingly restricted.

> First, in both cases, the scope of *account management* has been narrowly construed — even disregarding what other purposes (research, planning, etc.) might possibly be served. Thus, the loans management programme is limited to accounting for payments received against balances outstanding on loans. There appears to have been no anticipation of a need to analyse loan characteristics and actual repayment schedules, to forecast Corporation cash flow, to cope with different loan terms, etc. Similarly, in the case of the general accounting programme, there seems to have been no anticipation of a need (for example) to monitor income and expenditure against budgeted amounts or to relate general-ledger functions to other accounting activities such as payroll and inventory. Moreover, even though both loan-management and general-accounting activities were conceived of in similar terms, there appears to have been no requirement that the two programmes be linked on the computer.
>
> Secondly, in both cases, the scope of the *database* on which the above activities are based is also quite narrowly defined. The loans management database is limited to data on the loans themselves. Thus, there are no data on (for example) the loan beneficiaries (e.g., income, education or number of children) nor on the type of unit occupied (e.g., location, type of unit or floorspace). Similarly, the general accounting database is limited to data on the actual payment or receipt of cash. Thus, there are no data related to (for example) accounts payable and receivable or management of cash flow.

In short, identification of computer needs was something of a one-way street. Certain activities that were done manually are now being done by a computer, but they are being done in essentially the same way as they were done by hand. In other words, there seems to have been no attempt to see how the power and speed of a computer might enhance rather than just emulate the way in which things were done manually.

Turning now to the issue of how well the system meets the needs identified (even if, with hindsight, those needs now seem limited), the fundamental question is probably the wisdom of opting for custom programmes in a general-purpose, programming language (DIBOL) rather than customized applications of specialized off-the-shelf packages for database management or general accounting. Apparently, there were no substantial benefits in terms of price, performance or ease of use derived from choosing the first option. The disadvantages of custom programmes, on the other hand, are clear. A programmer is required for making even small changes in the programme, whereas users can learn to interact with the specialized packages. Indeed, the same package can be adapted to many different applications, whereas custom programmes are normally designed for one specific application. Programme packages, moreover, often represent many person-years of development (testing, debugging, optimizing, adding new features, etc.), whereas custom programmes inevitably do not. Put another way, the buyer of a custom programme pays for all of the development costs; the buyer of a package shares at least some of the costs with other users.

As for the programmes themselves, several unmet needs have been identified so far. For purposes of illustration, four specific shortcomings in the Loans Repayment Programme are discussed:

Cumbersome file structure

> During the life of a loan, its terms may change: for example, the monthly payment may be recalculated or the status of the unit may change from occupied to vacant or vice versa. However, the only way the programme can cope with such changes is to alter the terms of the original loan; the programme cannot keep records on more than one loan for a single housing unit. The Corporation needs a more flexible approach than this.
>
> The problem is that the programme database consists of two main files: a file of data on each of the approximately 17,000 loan accounts (about 2 Mb in size); and a file of data on each payment or other adjustment made to a loan account (which, at an estimated rate of 7,000 transactions per month, will grow to about 5 Mb in size over a year). If the first of these had been two files instead of one — one containing data on the beneficiary and the unit and the

other containing data on the status of the loan — more than one loan record could have been maintained for a single beneficiary/unit, including a "null" loan record corresponding to vacant occupancy. In this way, changes in the terms of a loan could be treated as a new loan supplanting but not erasing the previous one. In the meantime, as a short-term solution for managing records of vacant units, these could be indicated by a special code and the programme adjusted by addition of sub-routines in each sort/search routine to check for vacant units and treat them appropriately, plus any new routine(s) that might be specially required for managing vacant units.

Inflexible menus

One of the functions of the programme is to provide reports on data pertaining to selected records. For example, the programme will prepare a summary of the loan status for all units in a given project or a list of all transactions on a given account, but there are other reports the Corporation needs which the programme is not set up to provide. For example, the programme will provide a daily-transaction report on the day itself; but, once records are posted to the loan file, transactions can no longer be selectively retrieved on a chronological basis. In other cases, reports that appear on the screen cannot always be sent to the printer and vice versa.

The programme already contains routines for sorting, indexing and printing reports on either the screen or the printer, so it ought not to be difficult to alter the programme to provide a wide variety of reports and report formats.

Missing variables

The Corporation would like the computer to be able to print a list of accounts that are in arrears. Yet this turns out to be difficult to do, since there is no variable in the database for the due date of the next payment. Searching for accounts in which the date of last payment (variable LOLSDD) is more than a certain number of days past is not enough, as advance or lump-sum payments may mean an account is still in surplus, even though the last payment was made some time ago. There are ways of identifying accounts in arrears with the existing variables, but they are not entirely satisfactory. For example, as long as the monthly payment remains constant for the life of the account, the computer could search for accounts which have a current outstanding balance greater than it should be, given the amount and terms of the original loan and the time that has elapsed since repayment began, but this approach will become complex if any variation in the payment schedule occurs during the life of the loan (e.g., if the monthly payment is recalculated or temporary relief is granted in special cases).

Missing routines

According to procedures now in use, interest owing on loans is accumulated in a separate account. Repayment of this account does not fall due until the principal of the loan is fully paid off, whereupon payments are transferred to the interest account. Up to this point, no interest has been charged on the interest; now this changes. Interest is calculated on the unpaid balance and added to the account on a regular basis. However, the programme has no procedure for doing this; so, it is still done by hand.

In summary, the biggest constraint on use of the computer system is not hardware but software. Partly, this is due to a narrow conception of user needs and, partly, it is due to the decision to rely on custom programming instead of customized applications of generic accounting or database management packages.

Possible future uses

There is no shortage of potential computer applications in the Housing Corporation. By way of illustration, several generic programmes are discussed here, along with their possible uses by the Corporation:

Word processing

Almost any organization can benefit from word processing, if only from the improvement in the quality of the output, but word processing is particularly useful where: (a) the same document goes through many drafts; (b) the same letter goes to a number of different people with only small adjustments to each letter; or (c) the same paragraphs or sections are often required in different documents. It seems likely that the Corporation can find numerous instances of all three kinds of situations.

Database management

Using a computer for maintaining and processing records could help in planning and administration in the Corporation. For evaluating past projects and planning new ones, the Finance and Programmes Division and the Design and Planning Division would surely find it useful to have access to a comprehensive database on each unit, including socio-economic data on the occupants and design/location data on the unit itself. For the Administration Division, there are many kinds of records that could be maintained and processed more efficiently on a computer than by hand. Among these are: personnel records for approximately 250 employees; records of land assembled by the Corporation for future housing projects; operating and maintenance records for a fleet of some 70 vehicles belonging to the Corporation; stores records for office supplies and other items required by the Corporation.

Spreadsheet analysis

This kind of capability amounts to having a table of data that automatically recalculates itself whenever any of the numbers in the table are altered. By simply typing in labels, numbers and formulas, a user can adapt the table to meet whatever requirements he may have. The potential applications of such a capability are too numerous to mention in detail, but they extend all the way from engineering to economics and from finance to demography. To give just one example of how much can be done with a spreadsheet programme, consider the econometric model shown in Table 14, which can be run on any small personal computer.

Full-function accounting

Computers are often used to provide full-function accounting, often on a "one-write" basis. Typically, this involves not just the general-ledger function already implemented but also cash-flow management, payroll, accounts receivable/payable and inventory. The Corporation will have to determine just how far it wants to go on the road towards a fully integrated system, but it is clear that all of these accounting functions could be transferred to the computer.

Statistical analysis

For the Corporation, statistics is a basic tool for analysing housing wants and needs. The analysis of data on socio-economic, locational and environmental factors as well as interpretation of the results of questionnaires and other types of surveys forms an essential basis for planning and evaluating housing projects.

Table 14 Hypothetical macroeconomic simulation model for Jordan, 1980–6

	Year	*1* *1980*	*2* *1981*	*3* *1982*	*4* *1983*	*5* *1984*	*6* *1985*	*7* *1986*
			(Millions of 1980 Jordanian Dinars)					
Consumption	C	999	604	683	709	730	751	773
Investment	I	320	210	231	254	280	307	338
Govt Expenditure	G	518	543	571	599	629	660	693
Priv wage bill	W	200	387	524	565	596	628	662
Govt wage bill	W'	200	206	212	219	225	232	239
Profits	P	200	200	200	200	200	200	200
Taxes	T	225	564	548	578	617	659	704
Income net tax	Y	589	793	936	984	1022	1060	1101
Capital stock	K	500	710	941	1195	1475	1782	2120

Regression Coefficients:	-C-	-I-	-W-	-u-
	B0= 30.000	B4= 14.000	B8= -12.000	Note that u1, u2 & u3 are stochastic variables.
	B1= .720	B5= .780	B9= .240	
	B2= .550	B6= -.050	B10= .200	
	B3= .520	B7= .100	B11= .100	

Model Equations:

1. C(t) = B0 + B1*P(t) + B2*(W+W')(t) + B3*P(t-1) + u1
2. I(t) = B4 + B5*P(t) + B6*P(t-1) + B7*K(t-1) + u2
3. W(t) = B8 + B9*(Y+T-W')(t) + B10*(Y+T-W')(t-1) + B11*t + u3
4. T(t) = C(t) + I(t) + G(t) – Y(t)
5. Y(t) = W'(t) + W(t) + P(t)
6. K(t) = K(t-1) + I(t)

Notes and sources:

The model uses the Gauss-Seidel iterative method of solving complex systems of linear equations and is adapted from "Model I" in L.R. Klein and A.S. Goldberger, *An Econometric Model of the United States, 1929-1952* (New York: North Holland, 1955); see J.-H. Johansson, "Simultaneous Equations with Lotus 1-2-3", *BYTE* (February 1985), pp. 399-405.

The data for Jordan are taken from the National Planning Council, *Five Year Plan for Economic and Social Development, 1981-1985* (Amman: Royal Scientific Society Press, n.d.), Tables 1/1, 1/8, 1/9 and 1/10. Data for W, W', P and K are estimated for this Table. Note that G and W' have been assigned growth rates of 5 and 3 per cent respectively. Note also that P is assumed to be a constant. The regression coefficients are completely hypothetical and do not reflect any empirical data.

Project management

Project supervision from ground-breaking to occupancy is one of the principal functions of the Corporation. As projects become bigger and broader in scope (as with the Abu Nuseir New Town project), their management becomes correspondingly more complex. Appropriate computer programmes (using critical-path and programme evaluation and review techniques) can help the Corporation to monitor the performance of its contractors.

Computer-aided design (CAD)

Computers can help with the design of individual housing units as well as project sites. Architectural and engineering drawings can be created on the computer, moved about on the screen, and stored on disk for future use. In some systems, the computer can draw elevations and perspectives from plans. Similarly, computers can estimate the labour and materials required for a project and calculate a corresponding bill of materials. There can be little question that this is a highly promising area of potential computer use by the Housing Corporation.

Engineering calculations

Although the Corporation tends to rely largely on its consultants for engineering calculations related to its projects, there are times when staff members want to double-check some of the analysis done for them or to explore special issues such as seismic vulnerability.

Modelling and simulation

Although more sophisticated than most of what the Corporation is currently doing in the way of research and planning, computer modelling and simulation represent another potential application. Indeed, the computer will do a lot to help make such models accessible. Among the kinds of programmes that might be useful for the Corporation are models for assessing housing needs, predicting transportation impacts, forecasting population growth and distribution, and simulating the capacity of municipal drainage systems.

Thus, in summary, it should be clear that the question facing the Corporation is not whether there are additional applications for its computer system but, rather, which ones are most promising, given the configuration of the computer system and the cost of enhancing it. The first step in dealing with this issue is to examine the obstacles to further computerization in the Corporation.

The recommendations of this case are divided into four sequential phases. The first set of recommendations requires no decision or expenditure; they can be implemented immediately. The second phase entails a bit more in the way of personal and institutional commitment, but costs are still very modest. The third phase involves certain expenditures designed to enhance the computer system and make it into a user-friendly environment. The fourth and final set of recommendations provides for establishment of a comprehensive Corporation computer policy for implementation over the next several years.

The purpose of the first set of recommendations is essentially to raise the level of computer consciousness within the Corporation. This can begin immediately and could involve some of the following steps:

1. Contact vendors, users, training agencies, etc., to get whatever may be available in Arabic in terms of translations, summaries or other material related to the computer system. For example, the Arab Organization for Administrative Sciences in Amman has an active computer-training programme for computers similar to the one in the Housing Corporation.
2. Establish a "library" of books and magazines on computers and inform Corporation staff that they can "drop in" to the computer room, borrow the materials and try out the computer. Arrange one or two "demonstration programmes" to show what the computer can do and have a few "informal" applications (e.g., games or personal programmes) to show that computer use is not "all work and no play"!
3. Make contact with other computer users in related fields; find out what experience, advice, documents, programmes and other materials they have and might be willing to share.
4. Arrange presentations, demonstrations or videotapes on computers and computer uses for Corporation staff. There is a set of videotapes produced by Digital Equipment Corporation specifically for training on PDP-11 systems like the one in the Corporation. These videotapes are in English; but there may well be equivalent material in Arabic, at least for the basic operating system and utilities.
5. Start to monitor use of the present system, including the name of the programme used, the purposes for which it was used, the name of the user, the connect time and the CPU time. It may be

possible for the vendor to supply a computer programme that will do this automatically whenever users log on and log off the system; otherwise basic data can be gathered just by providing a logbook in which users sign in and sign out.

6. Continue to work with the vendor of the original system to remove the shortcomings in the Loans Management and General Accounting Programmes.

The purpose of the second phase is to begin the process of institutional commitment. Certain important decisions are to be made, but no actual expenditures would be involved. This phase can begin as soon as the activities of the first phase are under way. If technical assistance is secured from the United Nations or other sources, the timing of that assistance should be co-ordinated with this phase of activities.

1. Define a full-time position for someone, either in the Accounting Section or elsewhere, in respect of the computer system and its development. If this position can be filled by an expert recruited from the outside, well and good; but, if not, it is vital that the person selected should:
   (a) have the ability and the motivation to educate himself or herself about computers as quickly as possible;
   (b) have the skill and personality to be able to identify and make use of resources outside the Corporation; and
   (c) be released from other responsibilities to concentrate on development of the computer system.

   This person would be the natural counterpart for any international expert who might be provided to the Corporation.
2. Make a systematic study of potential computer applications in each of the Divisions, bearing in mind that computerization provides an opportunity not just to automate existing processes but also to refine and enhance these processes. It is not necessary to identify all possible computer applications, just some priorities for each Division.
3. Define some overall priorities for computerization in the Corporation. There is a wide range of potential applications, so choices will have to be made. The most sensible approach will probably be to choose applications with wide appeal (such as word processing, database management and spreadsheet analysis) even though these are sometimes regarded as "non-traditional" computer applications.

The purpose of the third phase is to make essential improvements to the computer system to make it attractive to potential users. This phase will involve certain expenditures on hardware and software.

1. Arrange to locate one or two of the existing four terminals to remote locations within the Corporation. Where the most appropriate locations may be will depend on the needs and priorities that have been identified in earlier phases.
2. Buy two to four more computer terminals. Unless Corporation needs and priorities indicate otherwise, it is unlikely that all or even any of these additional terminals will require on-line access to the main system. Accordingly, in view of the flexibility it would provide, purchase of stand-alone microcomputers rather than dumb terminals is recommended. A suggested configuration is: IBM-compatible CPU and operating system, bilingual keyboard, 256 Kb RAM and two floppy-disk drives.
3. Add a port or ports and provide software to allow microcomputers to communicate with the PDP-11 when desirable.
4. Buy more printers to meet needs different from those served by the present printer. For example, occasional users can use slow and cheap printers, but word processing needs better quality than the present printer provides. All printers should be fully bilingual; some, or perhaps all, of them could be moved to remote locations with the terminals.
5. Buy appropriate software packages. The choice here will depend on needs and priorities identified in earlier phases, but one of the most important factors to consider is how "friendly" the package is to the novice user. Thus, it may be wise (and economical) at this point to give priority to getting a number of popular packages for the microcomputers rather than one or two powerful packages for the PDP-11 (such as Datatrieve or Mini-tabs). More capacity can be added to the PDP-11 once users are ready to graduate to its sophisticated environment; for now, the priority is to provide easy access on a widespread basis. (An exception to this principle might arise in the event that priority is given to a full-function accounting package; this probably should be done directly on the PDP-11 and protected from unauthorized access.)
6. Arrange a simple but systematic training programme for novice users. Training should naturally be geared to the system and the priorities of the Corporation and should be in Arabic.

Finally, the purpose of the fourth phase is to initiate a long-term programme of development of a full-scale data-processing capability within the Corporation. As such, this phase should evolve naturally out of the experience of the previous ones. The recommendations made in the previous phases are all directed at the short term: steps to encourage Corporation staff to make use of computers. There should be no expectation that these steps will be sufficient to meet all possible computer needs for the next five years. In fact, in a way, it will be disappointing if the first three phases do not lead to demands for more facilities, for that will probably mean that the first phases have failed in their objective of promoting extensive use of the computer.

Case 17

Building institutional capacity (Sri Lanka)

There appears to be unanimous agreement on the need to strengthen the management capacity of urban local authorities in Sri Lanka. In two recent reports, both the government (Ministry of Local Government, Housing and Construction, no date) and the World Bank (June 1984) have concluded that urban local authorities in Sri Lanka are less and less able to discharge a more and more limited range of responsibilities. One method proposed for ameliorating the situation is the establishment of an urban information system.

There are presently 51 urban local authorities in Sri Lanka, 12 Municipal Councils and 39 Urban Councils. Together they contain some 70 per cent of the total population classified as urban in the Census and about 15 per cent of the population of the whole country. Both Municipal and Urban Councils are statutory bodies with the power to pass bylaws (subject to ratification by Parliament); to levy taxes, impose licences, charge fees and borrow money; and to spend money on the provision of certain statutory services. The latter consist principally of the supply of water, sanitation, solid waste disposal, drainage, local roads, street lighting, electricity, public markets, parks and playgrounds (Mendis, 1976). In general, Municipal Councils have somewhat wider powers than Urban Councils (e.g., in respect of taxation and staffing), but, in many other respects, they are similar. The 12 Municipal Councils, unchanged for the past 20 years, are Colombo, Badulla, Batticaloa, Dehiwela-Mt. Lavinia, Galle, Jaffna, Kandy, Kurunegala, Matale, Negombo, Nuwara Eliya and Ratnapura.

Local authorities in Sri Lanka have never been powerful compared to those in some other countries; for example, there has never been any local control over education or police. Never very strong, local authorities seem, over the past 30 years, to have fallen victim to a steady loss of power to sectoral agencies of the central government and a growing financial dependence on the central government for the responsibilities they retain. Moreover, the capacity of urban local authorities to manage even those services left to them seems to be diminishing. For example, according to figures cited in recent reports by both the government and the World Bank, the percentage of local expenditure actually financed from local revenues has declined from 44 per cent in 1976 to 29 per cent in 1979 and 26 per cent in 1982. By 1987, it was expected to have fallen to 18 per cent (World Bank, June 1984). It is tempting to blame local authorities for the situation they are in, and, certainly, there is ample evidence of their failings. Yet, it may also be that the process of increasing central control over local authorities has contributed to the very state of affairs it was meant to alleviate.

It is true that urban local authorities appear to have failed to discharge their responsibilities. According to both government and World Bank sources, urban local authorities exhibit many shortcomings. Many of them apparently still keep their accounts on a cash, not an accrual, basis. Many have no multi-year planning process and no capital budgeting. Many do not have any effective performance monitoring system. Virtually all are in arrears in the collection of revenues to which they are entitled; consequently, many have defaulted on payments for water and electricity. Few local authorities have planned maintenance schedules for urban facilities and infrastructure. In short, urban local authorities seem bereft of the most basic managerial skills and capacities (World Bank, June 1984).

However, these are not new problems. Tax-collection problems, for example, have been a feature of local government in Sri Lanka for at least 30 years (Mendis, 1976). As far back as 1960–1, the *Annual Report of the Commissioner of Local Government* complained that some councils were still collecting taxes on the basis of valuations made 15 or 20 years earlier — in some cases, prior to independence! The fact is that there are notorious political risks involved in trying to make changes to property taxation in any democratic country. Moreover, even though local authorities have since 1971 been required to reassess property values every five years, few have been able to do so because of the administrative costs involved. Ironically, local authorities cannot afford the expense necessary to make themselves rich! Changes in tax rates, moreover, require the explicit approval of the central government and must be justified in terms of an improvement in local services — which, ironically, is the very state of affairs for which higher rates are required. So, in a sense, local authorities are locked into a situation not altogether of their own making.

Until now, the central government seems to have been unwilling to bear either the political risk or the administrative expense of reforming the property tax. Instead, the central government has historically taken the expedient course of bailing out local authorities whenever they got into trouble. Apparently, this has been done in three ways: unconditional revenue grants, conditional capital grants and direct payments to utilities on behalf of defaulting local authorities. The combined total of these grants and subventions is about Rs 1 billion per annum and rising (World Bank, 1984).

Yet, the very nature of this approach — the lack of any specific criteria for allocation of grants, the crisis-management nature of the payments to utilities to cover defaults by local authorities, the lack of alternative sources of funds, the unpredictability of the whole process — makes it impossible for local authorities to plan effectively and may even appear to reward poor fiscal management on their part. As one Sri Lankan commentator (Leitan, 1979, p. 83) has said, the problems of local authorities are both cause and effect of their diminished role.

> Attitudes of administrators and of governments are also relevant in this connection. . . . [Some] officials tend to regard local government as largely redundant; for, they point out, [central] government departments already functioning in the field . . . perform . . . more efficiently, impartially and with less friction. The attitudes of governments . . . seem to be reflected in the reluctance to introduce any major changes which would associate local authorities as base-units in the development process. A combination of these attitudes (stemming at least partially from the inefficiency, corruption and indifference of local authorities themselves) has resulted not only in local authorities being bypassed but also in a steady erosion of their functions.

This conclusion is echoed by the World Bank, which summarizes one of the main themes of its last Sri Lanka country report (World Bank, 1984, p. v) by noting that everything points to the hypothesis that

> local authorities have lost a sense of responsibility for their own fiscal soundness and for financing their own urgently needed services, or, alternatively, that such a sense of responsibility is not being adequately supported by the current administrative and financial framework.

Ironically, in taking responsibilities away from local authorities and adding to the mechanisms for monitoring and regulating their diminishing responsibilities, the central government may have helped perpetuate the very situation it sought to correct. The World Bank concludes that, in the face of increasing control by central government and its own steadily declining powers, "local government morale . . . appears to be in the doldrums. . .".

One suggestion for strengthening the management capacity of urban local authorities is to establish a computerized information system. To this end, a "Management Information System for Urban Local

Authorities" has been proposed (World Bank, March 1984, Annex 2), whose purpose is:

(a) to help identify "local government urban service level deficiencies" in terms of "design characteristics" and "financial performance";
(b) "to provide a framework for the design and appraisal of a local government urban improvement programme"; and
(c) to help monitor "the performance of new assets against pre-set programmatic objectives and . . . the financial performance of each local government".

The data input to the proposed system consists of 109 primary variables (see Table 15). These variables are chosen to provide:

- "key global demographic, socio-economic and financial indicators";
- "a summary of existing sector performance" over eight sectors; and
- a detailed budget for the "last financial year for which information is available".

Naturally, there are many other indicators that will be derived from these primary variables, including income and expenditure totals, population growth rates, per capita performance and cost-recovery ratios. Furthermore, since coverage will extend to all 51 urban local authorities, it will be easy to compare results in one local authority with those in another or with the national average.

It seems doubtful, however, that the proposed design will provide information that is sufficiently detailed for effective monitoring and evaluation. First, the system is oriented primarily towards sectoral performance at the level of the local authority. There is no level of disaggregation below that of the whole local authority area. Secondly, the smallest unit of time appears to be a year. Thirdly, there is little or no distinction between "stocks" and "flows" — that is, between existing levels of service in each sector and the net additions that are due to be made under a particular programme or within a given period of time. The net effect of these characteristics is to make the proposed information system largely irrelevant for project management (at the local level) as opposed to programme monitoring and evaluation (at the central level). Moreover, as designed, the information system may contribute to the historical problem of poor morale at the local level.

By contrast, modifying the design of the system to make it a tool for project management (rather than programme evaluation) will make the system useful to urban local authorities, thereby strengthening their capacity to provide local services, providing an incentive to them to

Table 15 List of primary inputs for urban local authority information system (Sri Lanka)

No.	Variable	Units
	Basic Data	
1	Name of urban local authority	Name (characters)
2	Urban population, 1971	No. of people
3	Urban population, 1981	No. of people
4	Population density, 1981	No. of people per area
5	Average household income, 1981	Rupees per household
6	Population of urban poor, 1981	No. of low-income people
7	Labour force, 1971	No. of people economically active
8	Labour force, 1981	No. of people economically active
	Financial Summary	
9	Total capital expenditure, 1981	Rupees
10	Total recurrent expenditure, 1981	Rupees
11	Total internally generated revenue, 1981	Rupees
12	Total revenue assessment, 1981	Rupees
13	Total subventions and grants, 1981	Rupees
14	Total capital expenditure, 1982	Rupees
15	Total recurrent expenditure, 1982	Rupees
16	Total internally generated revenue, 1982	Rupees
17	Total revenue assessment, 1982	Rupees
18	Total subventions and grants, 1982	Rupees
19	Total capital expenditure, 1983	Rupees
20	Total recurrent expenditure, 1983	Rupees
21	Total internally generated revenue, 1983	Rupees
22	Total revenue assessment, 1983	Rupees
23	Total subventions and grants, 1983	Rupees
	Water Supply	
24	Sector performance	No. of gallons/day
25	Design indicator	Index
26	Total cost subject to cost recovery	Rupees
27	Direct cost recovery	Rupees
28	Indirect cost recovery	Rupees
29	Total target population	No. of people
30	Urban-poor target population	No. of people
31	Implementing agency(ies)	Code
32	Operating agency(ies)	Code
	Sanitation and Conservancy	
33	Sector performance (sanitation)	No. of sanitary latrines
34	Design indicator (sanitation)	Index
35	Sector performance (conservancy)	No. of metric tons
36	Design indicator (conservancy)	Index
37	Total cost subject to cost recovery	Rupees
38	Direct cost recovery	Rupees
39	Indirect cost recovery	Rupees
40	Total target population	No. of people
41	Urban-poor target population	No. of people
42	Implementing agency(ies)	Code
43	Operating agency(ies)	Code

Table 15 (cont'd)

No.	*Variable*	*Units*
	Drainage	
44	Sector performance (i)	Mean area water-logged
45	Sector performance (ii)	Mean duration in days
46	Design indicator (sanitation)	Index
47	Total cost subject to cost recovery	Rupees
48	Direct cost recovery	Rupees
49	Indirect cost recovery	Rupees
50	Total target population	No. of people
51	Urban-poor target population	No. of people
52	Implementing agency(ies)	Code
53	Operating agency(ies)	Code
	Roads	
54	Sector performance	Km of roads improved
55	Design indicator	Index
56	Total cost subject to cost recovery	Rupees
57	Direct cost recovery	Rupees
58	Indirect cost recovery	Rupees
59	Total target population	No. of people
60	Urban-poor target population	No. of people
61	Implementing agency(ies)	Code
62	Operating agency(ies)	Code
	Markets	
63	Sector performance	Sq m of covered space
64	Design indicator	Index
65	Total cost subject to cost recovery	Rupees
66	Direct cost recovery	Rupees
67	Indirect cost recovery	Rupees
68	Total target population	No. of people
69	Urban-poor target population	No. of people
70	Implementing agency(ies)	Code
71	Operating agency(ies)	Code
	Electricity	
72	Sector performance	No. customer connected
73	Total cost subject to cost recovery	Rupees
74	Direct cost recovery	Rupees
75	Indirect cost recovery	Rupees
76	Total target population	No. of people
77	Urban-poor target population	No. of people
78	Implementing agency(ies)	Code
79	Operating agency(ies)	Code
	Parks and Playgrounds	
80	Sector performance	No. of people served
81	Design indicator	Index
82	Total cost subject to cost recovery	Rupees
83	Direct cost recovery	Rupees
84	Indirect cost recovery	Rupees
85	Total target population	No. of people
86	Urban-poor target population	No. of people
87	Implementing agency(ies)	Code

Table 15 (cont'd)

No.	*Variable*	*Units*
88	Operating agency(ies)	Code
	Current Income and Expenditure	
89	Last reported financial year	Year
90	Income from rates	Rupees
91	Income from taxes	Rupees
92	Income from license duties	Rupees
93	Income from rents	Rupees
94	Income from electricity supply	Rupees
95	Income from water supply	Rupees
96	Income from other services	Rupees
97	Personal emoluments	Rupees
98	Travelling costs	Rupees
99	Supplies and requisites	Rupees
100	Repairs and maintenance of assets	Rupees
101	Transport, communication, utilities, etc.	Rupees
102	Interest payments, dividends and bonuses	Rupees
103	Grants, contributions and subsidies paid	Rupees
104	Pensions, retirement benefits/gratuities	Rupees
105	Equipment, land improvement, etc.	Rupees
106	Total revenue grants	Rupees
107	Total capital grants	Rupees
108	Total unpaid bills (prev 2 yrs) — water	Rupees
109	Total unpaid bills (prev 2 yrs) — power	Rupees

Source: Compiled from data in the World Bank Sector Mission *Report* in March 1984, Annex 2.

Notes:

Variables 2 and 3: If the boundaries of the urban area have changed between 1971 and 1981, they should be converted to the same standard.

Variables 5 and 6: Presumably these are measures of cash income only; the same units should be used for both variables (individuals and households).

Variables 7 and 8: Here again the boundaries of the urban area may have changed between 1971 and 1981. Presumably here too the definition of "economically active" means earning a cash income.

Variables 9–23: There is some potential duplication here with the data in variables 90 to 109 if or when the "last financial year for which data is available" (variable 89) happens to be 1981, 1982 or 1983.

Variables 24–32, etc.: Sectoral performance is assumed to be a measure of the magnitude of the services actually provided by a local authority within a specified period of time and which are subject to monitoring for the purposes of cost recovery. If information on either current overall or historical levels of service is required, appropriate data variables will have to be included. Data are entered in absolute terms rather than per capita or per-area terms, as the computer can readily calculate any required ratios.

The "design indicator" is assumed to be an index for describing the quality of the service provided; in some cases, it may be possible to use this as a weight in assessing sectoral performance.

It is unclear whether direct and indirect cost-recovery data are meant to be anticipated or actual data; either or both could be included.

Data on target populations may be difficult to get in some cases (e.g., roads, markets and recreational facilities). Names of the implementing agency(ies) and operating agency(ies) should be coded before entry into the database in order to save space.

Variables 44 and 45: It is unclear what is intended here, as per capita data are specified, but this seems like a reasonable interpretation.

keep it up to date and, at the same time, supplying useful data for performance monitoring at the national level. An information system without up-to-date information is as bad as no information system at all. Yet, poor data flow from local authorities to the central government has been a perennial source of concern. Even the constitutional requirement (under section 90.1) for local authorities to submit their accounts to the Auditor-General has not prevented extraordinary delays in data flow in the past.

The best way to make sure an information system is kept up to date is to make it useful for someone to do so. In other words, if the information system can be designed so that it is useful and accessible to local authorities, they will keep it current for their own benefit. Indeed, doing so will no longer be a chore carried out "for those people in Colombo"; it will be part and parcel of using the system. The only thing the central government will have to do is make sure that each local authority is using the same basic system — so that data from one local authority are consistent with data from another — and that a copy of the database (e.g., on a floppy disk) is forwarded periodically to Colombo.

This implies a somewhat different role for an information system than the one envisaged in the original proposal. Instead of merely an administrative convenience for the central government, the information system would become a key component of efforts to strengthen the managerial capacity of urban local authorities. Instead of being merely a record of urban performance, the information system would become an active component of urban development. As such, it would help reinforce other initiatives planned by the government to improve management in the urban local authorities.

Part five

Policy choices

Part five

Policy choices

The purpose of computerization is normally to improve individual and organizational performance. For present purposes, that means improving the management of human settlements. But the introduction of computers, as of any new technology, into human settlements agencies brings other changes too — some anticipated and some not. These changes affect the individual, the organization and the society.

At the level of the individual, computerization introduces a new element to the way people are evaluated, both formally and informally, by their colleagues. Suddenly, there is a whole new set of skills perceived as important — computer skills. People who have or who acquire these skills tend to get rewards and enjoy a status they might not otherwise have had. Established systems of evaluation based on background, formal education, seniority and even gender tend to get shaken up. Salaries increase quickly, promotion comes early, and peer status is raised for people who master the new skills. In short, a new technology upsets the existing "pecking order" in an organization. In the case of computers, this process tends to favour those with an analytical turn of mind, those most recently at school, those who are flexible enough to adapt to new ideas, those with the time and inclination to learn new skills, and so on.

In human settlements agencies, this will bring both good and bad results. For example, computerization may improve job and career opportunities for women and young people; but it may also devalue certain traditional skills, such as book-keeping or manual cartography, to the development of which individuals may have devoted years of their working lives. Similarly, when the demand for computer skills outruns the supply and because the skills are in any case highly portable, computers tend to promote staff mobility. For the individual that may mean a rewarding career; but, for the organization, it means high staff turnover and, in the case of government agencies, direct competition with the private sector and even with organizations in other countries.

Computers can also affect individual job-satisfaction. For one thing, computers can relieve human workers of tasks that are boring, tiring and repetitive. Computers can also help improve the accuracy, consistency and presentation of results. Finally, microcomputers with their user-friendly interface, allow and even encourage people to manage their own computing needs, without being dependent on specialists. This can help enhance even menial clerical and technical jobs and reduce feelings of alienation at professional levels.

Computerization also has broad implications for society as a whole. Some of these are obvious and widely publicized (though nonetheless refractory for that). There is the loss of personal privacy brought about by the greatly enhanced ability of computers to process existing data let alone store new data about individuals and their financial and social affairs. There are the problems of regulating the ownership of intellectual property, such as computer software, and of controlling transborder data flows. Computers have even created new opportunities for criminal activity such as fraud and embezzlement.

Other social implications are not obvious. For example, computers may help organizations improve services to their customers, but computers are pre-eminently impersonal, rule-following machines. This is their strength and their weakness. Thanks to the data-processing power of computers, management may be able to encompass greater variety than hitherto — but at the price of rigid conformity to underlying rules and regulations. For instance, computers can help make development-control procedures efficient and consistent, which should contribute to raising the standard of planning and lowering the costs of development; but there may be times when society is better served by delay or inconsistency. Computerization may well make the exercise of this kind of discretion more difficult than it was. Similarly, computer-assisted design and drafting (CADD) can mean public-housing schemes with dozens of different designs as easily as a handful; but the provision of so much variety may make real exceptions (such as self-help construction) that much less acceptable. Computers are, in short, perfect bureaucrats.

Computers also illustrate what Abraham Kaplan (1964) termed the "law of the hammer": if you give a small boy a hammer, he will soon find that everything he encounters needs pounding. Contemporary society is already probably the most researched and documented society in history. This is the era of data. We truly live in an information society — of which computers are undoubtedly both cause and effect, inspiration and manifestation. Yet it behoves us to remember that information in general and computers in particular are only a means to an end. While computers can do a lot (as we have seen in earlier sections of this book) to contribute to human settlements management, we must always be careful not to let the "joy of computing" divert us from its ultimate goals.

The third level at which computer technology can have unforeseen implications is that of the organization — which is the focus of the cases presented in this section. These cases examine three aspects of the relationship between computerization and human settlements policy, namely, the balance of power between different agencies responsible for human settlements; the balance between centralization and decentralization or between power at the centre and power at the periphery; and the nature and meaning of development itself.

The first case (Cyprus) presents a unique situation in which responsibility for human settlements planning and management is clearly divided between two communities. On the one hand, essentially technical matters, such as computerization, provide a useful and overtly apolitical opportunity for communication and co-operation between the north and the south. However, as the case shows, it is essential to preserve the existing balance of power, even when the available resources are relatively scarce. In the second case (Mexico), computerization is intended to help implement a policy of decentralizing urban population growth and development from Mexico City to a second tier of "middle-sized" cities. However, as the case shows, deliberate steps need to be taken to ensure that computerization has the desired, not the opposite, effect. Finally, the third case (Panama), where the existence of the Canal dominates everything, illustrates how computers can play a critical role in creating a desirable natural environment for human settlements.

Case 18: Maintaining the balance of power (Cyprus)

In Cyprus, there are two main authorities responsible for housing and town planning: one for the Greek-Cypriot community in the south and one for the Turkish-Cypriot community in the north. Each authority has a computer system — an NCR 9020 minicomputer in the south and two microcomputers in the north. So far the only fully operational application is a housing data bank running on one of the microcomputers in the north. Beyond this, there are plans for implementing a development control and monitoring system on the NCR computer, first in the south but eventually on an island-wide basis.

The case provides a picture of the problems and opportunities presented by the existing hardware, including the costs of data-entry, the constraints of data storage, the need for appropriate software, the opportunities for alternative outputs and additional uses of existing equipment. The case also draws attention to the fact that, as they now stand, the existing computer systems are almost totally incompatible with each other.

On this basis and bearing in mind the need to maintain the existing balance between the two agencies, the case proposes a joint workshop on the use of microcomputers for human settlements management as well as several immediate steps to help strengthen and integrate the existing computer systems in the two authorities.

Case 19: Computers and decentralization (Mexico)

In Mexico, the Secretaria de Desarrollo Urbano y Ecologia (SEDUE) is responsible for implementation of the Middle-sized Cities Project. The aim of this project is to channel demographic and economic growth into cities other than Mexico City, Guadalajara and Monterrey. One of the key elements in this project is a commitment to strengthening the processes and mechanisms for management at the local level. Computers and information systems are seen as one way of meeting this objective and several initiatives are already under way in SEDUE. Among these are a photo-interpretation and cartographic system, a minicomputer colour graphics system and a large-scale Urban Information System with approximately 100 remote terminals.

The case illustrates two frequent tendencies in relation to computerization in public agencies. First, the various computer systems have been allowed to develop more or less independently of one another, with the result that there is little that one system can do to reinforce use of the others. Secondly, although computerization is intended to promote decentralization by strengthening the periphery at the expense of the centre, the configuration and design of the computer systems (in particular, of the new Urban Information System) reflect an unremittingly centralist tendency. As the case suggests, computer technology is not inherently divisive or inherently centralizing; the effect of technology merely reflects the way it is implemented. So, if computer technology is to reshape an organization, implementation of the technology must be designed accordingly.

The case concludes by making three recommendations:

> First, microcomputer facilities both in SEDUE and in local planning offices need to be dramatically improved. Unless information can be processed at the local level, whether that information comes from a computerized system or some conventional source, computerization will have little impact on local planning and management capacity.

Secondly, communication with and down-loading of information from the IBM 4345 should be made easier.

Thirdly, plans and programmes should be developed for using all these resources to strengthen local capacities. This includes supplying planning and management units with appropriate hardware and off-the-shelf software, with technical support and service, and with access to suitable training and seminars.

Case 20: Computers and development (Panama)

In Panama, the Oficina de Planificación y Desarrollo del Area Canalera (OPDAC) of the Ministerio de Planificación y Política Económica is the chief governmental agency responsible for the management of human settlements in the Canal Area — which is to say the area that supports almost half the total population and three-quarters of the Gross Domestic Product (GDP) of the entire country. Central in every sense to the Canal Area is the Canal itself, and the staggering fact is that the operation of the Canal depends on a continuous and substantial supply of fresh water to the Canal — sufficient not only to cover losses due to evaporation and seepage as well as human and industrial consumption but also to replace the approximately 400 million litres lost to the Atlantic and Pacific Oceans for every one of the 10–15,000 ships passing through the Canal each year. Before anything else, therefore, human settlements planning in Panama means taking account of the natural environment.

To help meet its responsibilities for promoting the development of environmentally sensitive human settlements in the Canal Area, OPDAC proposes to establish a number of computerized information systems and environmental monitoring programmes. The purpose of the case presented here is to examine the relationship between information and the quality of life in human settlements.

The case makes two points. First, the case argues that one of the biggest obstacles to an effective use of information for human settlements planning in the Canal Area may be a conceptual approach that is too narrow. While it would be hard to overestimate the importance of the Canal to Panama, nevertheless the Canal is not the only thing that matters. In other words, maintenance of the Canal is a necessary but not a sufficient goal for planning in the Canal Area. The economic life of the Canal can be set at 20, 50 or 100 years, according to judgement; but the consequences of deforestation of the Canal Area — if that were indeed to occur — would be much more profound and enduring than that. Thus, it is important to preserve the Canal for as long as possible — but not to the exclusion of all other social and environmental planning objectives.

The second point illustrated in the case is the large number of interests that can often be involved in human settlements planning. In Panama, there are more than 30 different governmental agencies with a stake in the future of the Canal Area. Add to that the myriad private and voluntary groups, including land-owners, land-users, industrialists, financiers, labour unions, churches, student groups, house-owners, conservationists and farmers. With such a diverse institutional setting, the use and sharing of information is sure to be problematical.

With these points in mind, the case supports the idea that computers and information systems can play a key role in human settlements management — provided such systems are designed, implemented and managed with that end in view. For the Canal Area, therefore, the overwhelming need at this time is to clarify objectives, co-ordinate interests and formulate a plan. In conjunction with a development strategy that meets these needs, computerization will undoubtedly contribute to the design and implementation of appropriate projects and programmes for the improvement of human settlements in Panama.

Case 18

Maintaining the balance of power (Cyprus)

Computer systems have been installed in both the Department of Town Planning and Housing in the southern part of Cyprus and in the Offices of Town Planning and Housing in the northern part. The first has an NCR 9020 system, while the second has an Intertec Compustar Model 30 in the Office of Town Planning and a Sord M-23 in the Office of Housing.

The NCR 9020 computer is a 16-bit minicomputer built around a proprietary central processing unit running a proprietary operating system called IMOS (Interactive Multiprogramming Operating System). The system has 256 Kb of memory (RAM), which can be expanded to 512 Kb; and 81 Mb of disk storage (expandable internally to 162 Mb and externally to 324 Mb), 13.5 Mb of it removable. There is also a built-in cassette tape drive used primarily for the bootstrap programme. The system currently supports two terminals and two printers but can take up to 24 in total. The terminals are standard 12-inch CRT displays with 24 rows and 80 columns; they have separate numeric keypads but no function keys. Of the two printers, one is a high-speed line printer and the other is a letter-quality dot-matrix printer.

The principal programme available on the system is a custom-written COBOL programme for monitoring departmental processing of applications for development-control approval. The programme records details of each application, generates 35 different kinds of management reports and prints five different kinds of "form" letters. The programme requires three text "library" files to provide the basic input to the reports and "form" letters.

The programme is not yet operational, since authority to do so awaits a proposed new Planning Law, which is not expected until next year. In the meantime, the system is shut down in order to save the vendor's L230 per month maintenance charge. When it is operational, the programme is expected to have to handle about 15,000 applications per year, although only about 10 per cent of these cases are estimated to be active or pending at any particular point in time. The database is to consist of records each about 130 fields or 2 Kb in length. This means that the database could be as big as 30 Mb by the end of the first year — although it will be somewhat less than that if the file structure is sequential and variable-length fields are permitted. Currently active applications would thus take up about 3 Mb of file space.

The Intertec Compustar Model 30 is based on twin Z-80A microprocessors running at 4 MHz (although the User's Manual warns that DISKCOPY can take "5-10 minutes"). The Model 30 has 64 Kb of memory (RAM) and runs what it calls "M30 Standalone DOS 2.0 for CP/M 2.2". The Model 30 is equipped with two 700 Kb disk drives (5¼ inches) formatted for 80 tracks. (The User's Manual says 70 tracks but the local dealer insists it is 80.) The screen is 24 rows by 80 columns. The printer is also an Intertec model (99G): an 80-column, serial device with a 7x7 matrix printhead rated at a maximum speed of 180 characters per second. The computer requires 110 VAC and has an external transformer, while the printer runs directly off the mains (240 VAC). The Compustar has a full set of CP/M utilities (including ED.COM, ASM.COM and DDT.COM) as well as CSEDIT.COM for preparing and loading a secondary character set that makes use of the upper 128 (non-ASCII) characters and two programmes (TX.COM and RX.COM) for communicating with other Compustar computers. The system is equipped with MBASIC (version 5.21) and CBASIC as well as a "bare-bones" Wordstar package (version 3.0 without WINSTALL or Mailmerge).

The system was supplied about two years ago but has apparently not yet been used. The equipment is all in good working order, although the computer and printer had not been properly configured for each other. Although no specific uses are yet planned for the Compustar system, it is now ready for word processing (in English only) and programming in BASIC. Additional software (e.g., dBaseII for database management and Multiplan and Calcstar for spreadsheet analysis) are available from the local dealer. It may also be possible to "port" software from other CP/M machines over to the Compustar.

The Sord M-23 is an eight-bit machine based on the Z-80 microprocessor. The M-23 uses a proprietary operating system (FDOS and KDOS) and runs two 1 Mb floppy disks (5¼ inches) as well as two (expandable to three) 7.5 Mb hard disks. The system also has an SLP 160 dot-matrix printer with a wide carriage. The system has "bundled" software for word processing (Sord Word Processor) and integrated spreadsheet/database management (PIPS III), as well as Microsoft BASIC and various DOS utilities.

The main use of the Sord system to date has apparently been for establishment and operation of a Refugees' Housing Bank. This database consists of about 20 data fields (approximately 250 bytes) of data on refugee families and the houses allocated to them. The database currently contains about 2,000 records (families) and is expected to grow to 10, or even 20, times that size. Limited word processing is also carried out on the system. The local Sord dealer (in the southern part) is reported to have gone out of business; but there are several Sord users in Nicosia with whom contact could be made for advice and help.

In short, the authorities concerned with town planning and housing in Cyprus currently dispose of three different computer systems, one based on a minicomputer and the other two based on microcomputers. These resources reflect two weaknesses: they are not adequate for the needs and they are incompatible with each other.

The NCR system is to be used primarily for processing applications for development-control approval on a national basis. However, when it becomes operational, the system is likely to run into several difficulties:

Data-entry

Given the quantity of data that has to be keyed in for each application (i.e., the size of each record in the database), data-entry is going to be a huge undertaking. For one thing, data will probably have to be transcribed from the forms filled in by the applicant to "coding sheets" designed to facilitate keyboard entry. (It may be possible to bypass this step if the application form is redesigned with computer data-entry in mind; but some form of pre-processing or coding may still be necessary.) Then the data have to be typed into the computer terminal. (It is possible to "read" data into a computer electronically by means of a "character recognition" device, but this is not really feasible in the present circumstances.) Experience suggests that data-entry by unskilled staff may proceed at a rate of as few as four applications per hour; at that rate and assuming a total of 1,500 work-hours per year, data-entry alone will require three terminals, two full-time and one half-time. The system is currently supplied with only two terminals.

Skilled keyboard operators may be able to work at a faster pace, but this raises the question of whether data-entry is to occur in Nicosia or in the Department's five district offices. Given the desire to process applications as promptly as possible, it hardly makes sense to start off by waiting until the application can be forwarded to Nicosia and entered into the computer. Clearly, the wisest course of action would be to install a remote terminal or terminals in each of the district offices. However, this will mean the purchase of at least five more terminals as well as appropriate communication hardware and software. In that case, it may not be possible to provide skilled keyboard operators for all data-entry.

Data storage

Although storage requirements are relatively modest in terms of the active part of the database — applications "pending" are estimated to require about 3 Mb at any given time — cumulative storage requirements will grow rapidly. By the end of the first year, the database alone could be 20-30 Mb in size. Thus, it seems likely that the database will become too large for the available disk capacity sometime during the second or third year of operation.

The programme contains a menu option to erase records but there does not appear to be an option to archive "stale" records. This could be done by extracting these records from the master data file and "dumping" them to a separate file on the removable disk. However, this is not an altogether satisfactory solution in view of the cost of the removable disks and their relatively limited capacity (only 13.5 Mb, which translates into less than six months' worth of data). The common medium for data archiving on this scale is magnetic tape but that requires another type of drive which is quite expensive.

Software

The programme has had a limited test-run to date, involving entry and analysis of approximately 60 records, and was found (apparently) to be satisfactory. However, there was no attempt in this test to simulate "field" conditions or to "shadow" the manual system, in whole or in part (e.g., in one municipality). So the Department really does not yet know how the programme will perform in practice. It remains to be seen whether there is too much or too little information in the database, whether data fields are the right size, whether any new fields should be added or any existing fields deleted, whether the five "form" letters are sufficient for every contingency, whether the 35 management reports provide all the information which management needs, etc.

Because the programme is custom written, any changes in the structure or use of the database (such as those proposed above) will require reprogramming (in COBOL). Furthermore, the programme lacks the flexibility of a full-feature database management system. For example, its search and print routines are all rigidly predetermined; and there is no built-in sort routine at all. Apparently, searching can be done on the Application Number or the Plot Number, but there is (currently) no menu option for finding records on the basis of, say, a name or address or for listing all records pertaining to, say, plots of more than one hectare or buildings of more than one storey. Similarly, there are 35 reports that can be produced, but there is no menu option for adding (for example) percentage calculations to the printouts.

Printer output

The letter-quality printer is to be used to generate form letters required for each application. It should be noted that if applications were to require an average of seven letters each, the printer will have to produce over 100,000 letters per year — which means about 500 letters per day. That is fairly heavy-duty use for a letter-quality printer.

The high-speed printer is to be used primarily for printing management reports. Given the relative infrequency with which these reports are likely to be needed (e.g., monthly or even weekly), it seems likely that this printer will be idle a good part of the time.

Both form letters and management reports are required to be printed in Greek. However, neither printer is capable of printing the lower-case Greek characters. Thus, all printer output will be in upper-case only. If this is the case and in view of the likely quantity of form letters, it may be wise to do all printing on the high-speed printer, even if the print quality is not as good.

Once again, however, there is the question of whether this activity is to be centralized in Nicosia or decentralized to the five district offices. Perhaps form letters could all be printed and posted from Nicosia, but, clearly, it would be helpful for district offices to be able to get management reports directly from the computer — to say nothing of being able to answer the "What's happened to my application?" types of inquiries from members of the public. That means either installing a printer in each of the district offices or modifying the report routines to output to the screen as well as the printer.

Other uses

Given the present configuration of the system and its intended application, there is not a great deal of excess capacity that could be made available to other uses.

For one thing, the present hardware will probably soon be working at close to capacity (apart from one of the printers). If significant disk and magnetic-tape storage are added to the system, additional uses (such as a housing database for the Housing Management Section) could be accommodated in the system for no more than the price of the additional terminals, printers, communication network and software required; but the question of whether this would be the most cost-effective solution or whether it might be wise to invest in an alternative system would have to be determined on a case-by-case basis. For graphics-intensive uses in particular, it would probably be wise to consider an alternative system.

In general, it should be borne in mind that the NCR 9020 is not inherently a very flexible system. A COBOL compiler and a BASIC interpreter are available for the 9020 system, as well as a number of custom-written, special-purpose applications programmes; but there is little else. The word processing programme is (by the NCR dealer's own admission) "unsatisfactory". Moreover, according to the same source, there is no relational database management system and no spreadsheet programme available for the 9020. In short, as the sales brochure candidly acknowledges, "the NCR I-9020 is an interactive, transaction-oriented system for traditional accounting systems. . .". Other kinds of applications (such as graphics or mathematical forecasting) are not going to be easy to do.

The Intertec system is not as big or powerful as the NCR system but it is more flexible. This means that varied uses can be contemplated for the Compustar, although perhaps not on the same scale as with the NCR system. The most immediate and obvious application is for word processing (although there may be better software packages for Turkish and Greek than Wordstar). The existing programme could be enhanced in two ways: (a) by "installing" the programme properly for the printer (which has most of the typical features of a dot-matrix printer, including enlarged and condensed pitch, superscripts and subscripts), and (b) by adding Mailmerge and other supplementary programmes. The first would mean either getting a programme called WINSTALL.COM that does the installation automatically or getting a table of the appropriate addresses for Wordstar 3.0 and customizing them with a debug programme (such as DDT, which the Compustar already has). For all these, the wisest course would be to check with the original supplier of the Wordstar package to find out why the other programmes were not supplied originally and how they can be obtained now.

Other commercial packages for the Compustar are available from the local dealer, including industry-standard programmes for database management (e.g., dBaseII) and spreadsheet analysis (e.g., Multiplan or Calcstar), and others. Commercial software for the Compustar may also be available by mail order from sources in the USA or elsewhere. Programmes such as these would make it possible to use the microcomputer for a housing database, a simple development-control monitoring system, population forecasting, economic modelling, statistical analysis, budgeting, etc.

A third set of applications could derive from the large quantity of public-domain software that is available for CP/M-based microcomputers. (Public-domain software means programmes which are no longer protected by copyright and so can be freely given away.) Not all such software will be able to work on the Compustar, but some programmes will. Certainly, programmes written in Microsoft MBASIC ought to run on the Compustar, providing that no machine calls are involved. Similarly, the Compustar has its own CBASIC compilers (including one especially for 2.02 and 2.04 source code); so these programmes too should run on the Compustar.

However, all these programmes will first have to be transferred to the Compustar. There are two ways of doing this. The easiest way is to use a special programme (such as Media Master or UNIFORM) that can copy from one disk format to another. Of course, this depends on having such a programme that will run on the source microcomputer and that can write the Compustar format. Failing that, the alternative is to connect the source microcomputer directly (or via modem) to the Compustar and to "port" the programme across to the Compustar (which can then write the programme in its own disk format). This method is slow and cumbersome and requires special communications software for one or both machines.

The Sord system already has one fully operational application (the Refugee Housing Data Bank) under way. This use appears well within the capabilities of the hardware and software, although hardware and software may prove constraining as the size of the database grows. Opportunities for additional uses of the system would seem good — subject, of course, to there being sufficient time available on the keyboard. The software available (PIPS III) is not a widely used package (outside Sord machines) but it seems to have the essential features of spreadsheet programme (including primitive database management functions such as finding and sorting). On the other hand, it lacks the features of a so-called relational database management system (such as, say, dBase); these features may become important as use of the system expands. Similarly, the Sord word processing programme is not a standard one (such as Wordstar, for example) but seems to contain most of the essential features (although its capacity to display Turkish or Greek characters on the screen or printer is not known). As in the case of the Compustar, the Sord is equipped with Microsoft MBASIC, thus giving access to programmes written in that language.

Thus, the three computer systems examined here present a number of different kinds of problems and opportunities. Strictly speaking, only one computer application to planning is currently in operation; and that is the Housing Data Bank running on the Sord system. An ambitious scheme — a sophisticated management system for an estimated 15,000 development-control applications per year — awaits implementation on the NCR system, as soon as there is statutory authority for it. Development of other computer applications is possible on all three systems. However, this will probably prove less easy on the large system than on the two small ones, even though the financial incentive to find other applications might be felt to be the greater in the case of the large system.

Unfortunately, as the table below indicates, the three computer systems are about as incompatible with each other as they could be:

	NCR 9020	*Compustar*	*Sord*
Central processing unit	Proprietary	Twin Z-80A	Z-80
Data path	16-bit	8-bit	8-bit
Floppy disk capacity	None	700 Kb	1 Mb
Operating system	IMOS	CP/M 2.2	FDOS/KDOS

In fact, as matters stand, about the only feature shared by the three systems is that they can all run programmes written in BASIC. Thus, if a BASIC programme can be physically transferred from one system to another, it should be able to run on that system too.

However, even this is subject to qualification because: (a) the NCR system uses a proprietary interpreter; and (b) while the Compustar and Sord systems both use interpreters written by Microsoft, there may still be differences between them. What this means in practice is that more or less the same BASIC programme will run on all three computers, but there may have to be modifications to things such as screen and printer routines or the treatment of parameters in certain functions in order to make a programme from one system run on another.

The limited and incompatible nature of available computer resources plus the need to maintain a balance of capability between the various authorities concerned makes it difficult not to propose additional investment in equipment. Subject to the constraints of existing resources, however, the opportunity probably lies in the area of training for and assistance with implementation of a broad range of basic applications.

As there is little other than the BASIC programming language that is common to all three computer systems, one possibility would be for BASIC to form the "core" of a workshop or training session for intended users. However, learning to programme in BASIC is only indirectly going to make people good planners. Moreover, BASIC programming is unlikely to become a common activity on any of the existing computer systems. There may well be good reasons for learning BASIC and there may be lots of useful planning principles and techniques that could be taught along with BASIC, but, in terms of meeting job-related user needs, teaching BASIC programming would clearly not give very satisfactory results.

Another possible solution would be to find a planning programme written in BASIC (so that it would, in theory at least, run on all three systems) and build the whole training programme around it. In fact, the United Nations Centre for Human Settlements (Habitat) already has a piece of software that would suit the purpose, namely, its Urban Data Management Software (UDMS). However, the practical difficulties involved in transferring and adapting the programme to each of the three computer systems may be considerable. It is difficult to estimate how much time this would take without knowing in detail what difficulties the programme would encounter on each system; but the work would in any case depend on a programmer having access to the systems on which the programme is to run (i.e., an NCR 9020, an Intertec Compustar Model 30 and a Sord M-23). In the end, the expense and effort required would not likely be justified by the relatively limited use to which the programme would be put.

Instead, it is recommended that training should proceed from the following assumptions:

1. It will probably be necessary to expand the NCR 9020 system by adding some five remote terminals, one in each of the district offices. Suppose that, instead of buying "dumb" terminals that could do nothing but communicate with the main computer, the Department were to buy "smart" IBM-compatible microcomputers that could both run programmes of their own and serve as remote terminals. With appropriate software — such as Wordstar for word processing, dBaseII for database management and Supercalc for spreadsheet analysis — these machines could serve not just for remote data-entry but also (on a stand-alone basis) for a variety of other planning applications at the district level.

2. The use of the Intertec Compustar system for planning purposes is severely constrained by the lack of software. Suppose that funds were made available for the purchase of, say, two commercial software packages such as dBaseII for database management and Supercalc for spreadsheet analysis. Add those to its existing word-processing capability (Wordstar) and the

Compustar system would be capable of many (and varied) planning applications, all of them using the same software as the "smart" terminals on the NCR system.

3. The use of the Sord system for planning purposes is the most advanced of the three at this time. The system already has the hardware and software necessary for word processing, spreadsheet analysis and (limited) database management. However, although similar in function to the programmes referred to above, the software currently in use on the Sord system uses different commands. Suppose that funds were made available to provide the same software for the Sord as could be running on the other two systems. Each system would need its own distinct versions of the software, but they would all look the same to users — the same commands, the same capabilities and limitations, the same data file structures, etc.

The essence of these proposals is to help enhance, at relatively modest cost, the flexibility and applicability of each computer system in relation to planning needs, but a byproduct of these actions would be to provide a broad "common core" around which to build a training programme for users of the three systems. Instead of focusing on BASIC computer programming *per se*, training could focus directly and entirely on computer applications to planning. In these circumstances, it would be quite feasible to contemplate including in the training programme some of the following types of computer applications (subject, of course, to participants' interests and preferences):

- creating a record of decisions, minutes, etc. which can be searched at will for those which deal with some particular matter;
- creating a database on housing, commercial establishments, historical monuments, road signs, etc., searching it on any field and printing user-defined reports;
- comparing the results of and assumptions in half a dozen different techniques for projecting data on small-area population;
- practising basic techniques of statistical analysis as applied to typical planning problems;
- preparing an investment programme under various assumptions as to costs and benefits, performing sensitivity analyses on key variables and calculating internal rates of return;
- carrying out simple exercises in structural analysis, mapping and computer-aided design; and even
- demonstrating the Urban Data Management Software (UDMS) referred to above that would be so difficult to convert to run on all three systems.

To summarize: while a training programme could be designed around BASIC programming within the framework of existing resources, it would be far from satisfactory. Since those existing resources will in any case have to be augmented, at least to a modest extent, it makes sense to design a training programme in that enlarged context. The result would be a more attractive programme. Naturally, it would be helpful if the training programme could also be co-ordinated with purchase of the microcomputers to serve as remote terminals, so they could conveniently be used for training while they are all in one place and before they are distributed to the district offices.

Equally important and closely connected with training is the question of implementation. The main purpose of training planners about computer applications is to show, largely at a conceptual level, how computers can help them do their jobs effectively and efficiently. The object is not to teach planners how to create programmes so much as how to use them. Some planners will also develop some programming skills as time goes by; but initially, at least, they will need help in implementing appropriate applications.

Advice and assistance with implementation of computer applications is particularly important in the present case, since progress to date is (as indicated above) still fairly modest. There is not yet the range or extent of experience of computer applications to town planning and housing in Cyprus that can provide the basis for complete self-reliance. On the other hand, technical assistance in this regard need not be long-term. The important thing is to make sure that planners can see at least one good application running smoothly on their system, that they really understand what the computer is doing and how it is doing it, and that they start to share with users of the other systems the experience and knowledge that are gained thereby.

Thus, the critical feature of technical assistance in relation to the implementation of computer applications is not so much the extent of the assistance as (a) its timing in relation to the training programme and (b) the fact that it is provided at all.

Case 19

Computers and decentralization (Mexico)

The pattern of human settlements in Mexico is dominated by the process of urbanization. "The process of urban development", according to the current national urban development plan (the *Programa Nacional de Desarrollo Urbano y Vivienda, 1984–1988*), "has been one of the transcending factors in our social and economic evolution, particularly in the present century."

The details of this process are summarized in Table 16. In 1940, four out of five Mexicans lived in rural areas (i.e., in settlements of fewer than 15,000 people); today (1980) more than half the population lives in cities. More than a quarter of the total population of the country is concentrated in the largest three cities of Mexico — Mexico City, Guadalajara and Monterrey. The average annual population growth rate for the period between 1940 and 1980 has been much higher in the urban areas (especially the large and middle-sized cities) than in the rural areas. By 1990, the total population may exceed 85 million and three out of five Mexicans will live in urban areas, half of them in the largest three cities. By the year 2000, the population may be well over 100 million; urbanization may have slowed somewhat but nearly a third of the total population of Mexico could be concentrated in the largest three cities.

The principal element in the current national urban development plan is the *Proyecto Estratégico de Ciudades Medias* (the Middle-sized Cities Project). There are 59 middle-sized cities in Mexico, with populations ranging from 100,000 to under 1 million.

> The objective of the Project is to channel the concentration of population and economic activities into the middle-sized cities and in so doing to make them into a real alternative to the large cities.

Specifically, the goals of the project are described in the plan in the following terms:

> By the year 2000, the Project expects to have redistributed to the middle-sized cities four million inhabitants who would otherwise

> have migrated to the metropolitan areas. . . . 2.9 million to Mexico, 0.6 million to Guadalajara and 0.5 million to Monterrey — which represents eleven per cent of the total population that will be drawn to these cities [between now and the year 2000].

Implementation of the Middle-sized Cities Project is to entail a broad array of initiatives designed to improve the quality of life in these centres. Among the various activities contemplated are projects to improve urban infrastructure and urban services, plans to establish land banks throughout the country and measures to preserve and protect the nature environment. Perhaps most significant of all, however, is the government's commitment in the national urban development plan to "strengthening the processes and mechanisms for planning and management of urban development at the local level, including questions of finance and public participation. . . " (p. 150).

As the agency responsible for implementation of the national urban development plan, the Secretaria de Desarrollo Urbano y Ecologia (SEDUE) recognizes the importance of good information for planning, both for its own purposes and as a means of "strengthening the processes and mechanisms for planning and management. . . at the local level". In fact, there is a great deal of planning information already available. To give just one example, SEDUE's predecessor, the Secretaria de Asentamientos Humanos y Obras Públicas (SAHOP), carried out more than 10,000 separate studies related to human settlements over the six years from 1977 to 1982! Moreover, details of all these studies are readily available in a three-volume catalogue. So there is no basic shortage of information.

The problem is rather one of getting the right information to the right person at the right time. Accordingly, SEDUE is moving to the use of computers as a means of improving the supply of information for human settlements planning and management. At least four different systems are currently in use. These are:

1. A Franklin Ace 100 microcomputer (without a printer), with BASIC as well as software for spreadsheet calculation and database management, which is located in and used by the Subdirección de Programación y Evaluación of the Dirección General de Desarrollo Urbano (DGDU).
2. A Sci-Tex/Hewlett Packard colour graphics system serving the whole of SEDUE.
3. A Data General supported system for photo-interpretation and cartography operated by the Dirección del Sistema Nacional de Información para el Desarrollo Urbano (DSNIDU) and serving the whole of SEDUE.

Table 16 Distribution of human settlements in Mexico by size, 1940–80

	1940	*1950*	*1960*	*1970*	*1980*	
			Number of Settlements			
Large cities	1	1	1	3	3	
Middle-sized cities	5	10	16	32	46	
Small cities	49	73	106	143	218	
Urban-rural centres	631	824	1,089	1,474	1,767	
Rural centres	104,802	97,417	87,793	94,254	122,679	
Total	105,488	98,325	89,005	95,906	124,713	
			Total Population (000's)			Annual Growth Rate
Large cities	1,560	2,872	5,210	11,611	17,508	6.04%
Middle-sized cities	781	1,927	4,059	6,766	11,038	6.62%
Small cities	1,587	2,410	3,778	4,747	6,667	3.59%
Urban-rural centres	1,973	3,940	5,138	7,514	9,257	3.86%
Rural centres	13,748	14,630	16,738	20,058	22,376	1.22%
Total	19,649	25,779	34,923	50,696	66,846	3.06%
			Population as a Percentage of Total			
Large cities	7.94%	11.14%	14.92%	22.90%	26.19%	
Middle-sized cities	3.97%	7.48%	11.62%	13.35%	16.51%	
Small cities	8.08%	9.35%	10.82%	9.36%	9.97%	
Urban-rural centres	10.04%	15.28%	14.71%	14.82%	13.85%	
Rural centres	69.97%	56.75%	47.93%	39.57%	33.47%	

Source: Adapted from data in the ***Programa Nacional de Desarrollo y Vivienda, 1984–88***, Table 1, p. 23.

Note: Large cities are cities with a population of 1,000,000 or more.
Middle-sized cities have from 100,000 to 999,999 inhabitants.
Small cities have from 15,000 to 99,999 inhabitants.
Urban-rural centres have from 2,500 to 14,999 inhabitants.
Rural centres have fewer than 2,500 inhabitants.

4. An IBM 4345 running MANTIS — a management-information system package — with (apparently) some 100 remote terminals and printers operated by the Dirección General de Organización y Sistemas (DGOS) for the whole of SEDUE.

The Franklin Ace microcomputer system is used for both planning and management purposes. For the first, programmes have been written in BASIC (a) to provide various estimates of demographic growth and (b) to apply the Hansen gravity model to forecasts of the interurban distribution of population. For the second, a database management programme has been written (also in BASIC) to monitor the progress of some 50 studies and projects currently being carried out by the DGDU. The system is clearly being put to good use and could benefit from access to generic spreadsheet and database management software; however, one microcomputer is hardly adequate for the needs of DGDU. Furthermore, the Franklin Ace 100 itself is fairly limited in comparison to the current generation of equipment in terms of speed, memory size and disk capacity.

The Sci-Tex graphics system consists of a dedicated minicomputer, mass storage devices, digitizer, scanner, plotter and other peripheral devices. It produces simple maps, charts and other graphics in colour.

The photo-interpretation and cartographic system is based on Data General Dasher and Eclipse minicomputers. They are used for production of maps and architectural drawings from aerial and terrestrial photographs supplemented by site visits. For example, mapping of all 59 middle-sized cities is complete on the basis of 1980 photography and about half complete on the basis of 1982 photography. With the computer system, maps are drawn to scales of 1:10,000 and 1:20,000; maps to scales of 1:15,000 and 1:25,000 can also be produced if desired. The system also has the ability to produce customized overlay maps (such as land suitability maps) on demand. Using low-level air photography (from 700 m above ground) and site visits, the system can also produce planning maps to scales of 1:1000 and 1:2000; coverage at this scale is, of course, far from complete. The main limitation of the system is that it is based almost entirely on photography and thus records only what is visible. This means that less tangible social and economic factors such as land values or employment characteristics do not appear on its maps and are not included in its databases.

Finally, the mainframe Sistema de Información y Seguimiento para el Desarrollo Urbano (SISDU) is as ambitious as the microcomputer system is modest. Its aim is to provide a comprehensive database — 36 different "screens" of information accessible through 11 different menus — on each of the 59 middle-sized cities of Mexico. Eventually, the system will be expanded to include the three large cities and a third tier of 106 "growth centres" (*centros de apoyo*) which, together with the middle-sized cities, constitute the entire Sistema Urbano Nacional. For each of these cities, the database will contain data on their environmental, physical and social characteristics. The system manual classifies the contents of the system in the following terms:

1. General Data
2. Demographic Aspects
3. Economic Data
4. Infrastructure
5. Urban Facilities
6. Current Land Use and Housing Stock
7. Status of Local Planning
8. SEDUE Programme Monitoring
9. SEDUE Project Monitoring
10. Other Federal and Local Government Activities related to Urban Development

Altogether there must be some 500 different data fields provided for each city.

The very scale of the undertaking makes it all the more important to assess how effective the system can be in meeting the needs of DGDU. For large quantities of data do not necessarily make a good information system. Information systems are essentially tools for decision-making. They are not repositories or archives; so they do not need to contain (as does the *New York Times*) "All the news that's fit to print." Information systems are not meant to replace other sources of data (statistical reports, atlases, yearbooks, etc.) but to supplement them. The purpose of information systems is to provide efficient access to data relevant for planning and management. To this end, the following criteria are proposed, not with a view to evaluating what the present system can or cannot do, but rather as a means of describing what features are likely to be important in making it useful to DGDU.

Level of aggregation:

Some planners may feel that data on a city-wide scale are too general for planning purposes. They may want data at the level of the neighbourhood or some other suburban unit.

Scope of the data:

Some planners may want to add data to and delete data from the information system. As a general rule, the data in a computerized information system should be data that are subject to fairly frequent change (because data can be changed very easily in a computer) and/or data that are useful for manipulating in some way (because computers are particularly good at manipulating data). Data which do not change over time and data which are included "just for the record" can as well be stored "off-line" in hardcopy form. Thus, for example, some planners may feel that they do not need data on latitude and longitude in the information system (because they can as easily look it up in a book when they need it) but they do need up-to-date data on air pollution and water quality.

Nature of the data:

Some planners may prefer raw data to the categories and norms built in to the information system. For example, they may want to know how many hospital beds or how many classrooms there are in each city rather than whether each city does or does not meet some norm.

Data timeliness:

Some planners may need to know the date of the data presented to them by the information system. They may also want some assurance that there is an established process for regularly updating the information, so they can be confident that they are getting the best available data.

Time series:

Some planners may need historical data as well as up-to-date data, so that they can analyse trends and flows as well as states and conditions.

Screen layout:

Some planners may find the display screens too verbose, rigid and poorly laid out for some purposes (e.g., rapid scanning). For example, they may want to see a lot of information in a concise form on the screen at one time or they may want to combine information on different characteristics or for different cities on a single screen.

Hardcopy format:

Here, much the same considerations will apply as for screen layouts. That is, some planners may want to be able to adjust the format of their hardcopy printout according to their particular needs.

Data coding:

While there is clearly a balance to be struck between verbosity and obscurity, some degree of data "coding" saves space and reduces errors. For example, coding a characteristic such as "commercial and industrial centre" as "com-ind" preserves the meaning while at the same time saving a lot of space and making it unlikely that someone will make a spelling mistake or enter "industrial" before "commercial" — both of which would cause the computer to think it was a different category from the one intended.

Data checking:

Some planners may want built-in data checking to ensure that values fall within known ranges (e.g., no temperatures above 50 degrees), that elements add to totals (e.g., all population distributions add to total population), that apparent inconsistencies are flagged (e.g., labour force larger than total population), etc.

System interrogation:

Some planners may want to be able to bypass the menus provided in the system and manipulate the data themselves directly on an interactive basis.

Data-processing:

Some planners may want to be able to process the data in the information system. For example, they may want to identify cities having certain characteristics (e.g., cities where the proportion of school-age children is above a certain level), or to do statistical analyses of urban characteristics (e.g., correlations between population density and the quality of various urban services), or to test various planning models (e.g., gravity models for estimating migration patterns or travel demand between cities).

On balance, therefore, a lot of factors are involved in the design of a human settlements information system. Not all of them will be equally important in any given case, and trade-offs will undoubtedly have to be made in the face of limited resources; but the fundamental point is that an information system has to be useful. The quantity of data in an information system matters little, unless they can easily and usefully be put to practical use.

On the basis of the foregoing and bearing in mind the need to focus on short-term and pragmatic recommendations, there are three main suggestions to be made:

Strengthen the microcomputer system

Compared with the level of investment in the other computer systems, microcomputer facilities in DGDU are seriously underdeveloped. Thus, the number-one priority for DGDU should be to replace its microcomputer with at least a dozen microcomputers and printers and to encourage local planning authorities in the middle-sized cities to equip themselves in a like manner.

To people unfamiliar with computers, it may seem wasteful to spend money on microcomputers while there is spare capacity elsewhere, but microcomputers have some unique advantages. One of these is the quantity of user-friendly software to which they give access. For microcomputers there are generic packages for functions such as word processing, database management, spreadsheet analysis, statistics, project management, graphics and modelling which still have no equal on large computers in terms of flexibility and ease of use — to say nothing of cost. Moreover, there are dozens and even hundreds of microcomputer programmes specifically related to planning in the fields of land use, housing, transportation, education, health care, infrastructure, etc. Many of these special-purpose programmes are distributed by enthusiasts (including the United Nations Centre for Human Settlements) for no more than the price of a floppy disk.

There are other advantages to microcomputers. As a general rule, for a given configuration, microcomputers are cheaper to buy, operate and maintain than large ones. Microcomputers need no special facilities or premises to operate in. If the central processing unit (CPU) in a time-sharing system fails, the effects are catastrophic: everyone has to stop work. If a microcomputer CPU fails, that machine is out of operation, but no other machines are affected.

For remote operation — such as in each of the 59 middle-sized cities — microcomputers can work independently. They do not need any direct on-line link to a central computer. That link can be provided by telephone if desired or by the simple expedient of mailing a disk. Little of the information that planners normally use is so timely or so critical that it must be transmitted in seconds or minutes; most of what is valid today will still be valid a week later.

In addition to current applications, there are many potential computer uses in DGDU that are within the capabilities of a microcomputer. For example,

- plan-making: modelling and statistical analysis of data on issues such as land use, population, housing, transportation, industrial development and capital investment, etc.;
- plan-monitoring: database project management applications to help DGDU monitor and evaluate progress towards implementation of the Middle-sized Cities Project; and

- planning support: information, training and software for local planning authorities to strengthen their ability to manage the development of their own cities.

Facilitate microcomputer-IBM 4345 communication

Some planners may even find it easier to use the SISDU information system on a microcomputer than on the IBM 4345. If the basic database of 59 records contains, say, five kilobytes in each record, then the whole thing could be stored on a single floppy disk and processed with a standard microcomputer database programme. The only problem is getting the database out of the IBM 4345 and into the microcomputer.

Accordingly, steps should be taken to facilitate communication between microcomputers and the IBM 4345. Ideally, the large computer should be able to read and write floppy disks that can be read and written by microcomputers. Alternatively, a modem link should be provided. On the software side, either or both computer systems should be equipped with software to facilitate file conversion from one environment to the other.

Use microcomputers to strengthen local planning capacity

One of the principal objectives of the Middle-sized Cities Project is to strengthen the planning and management capabilities of the local authorities. Microcomputers provide a low-cost mechanism for the centre not just to plan and monitor how this can and should be done but also to help make it happen.

If computers can enhance the ability of DGDU to plan and manage human settlements on a national basis, there is no reason to suppose that they cannot do the same for local authorities on a local scale. The only way to supply computer capacity to local authorities on an affordable basis is by resorting to microcomputers.

It goes without saying that computerized local authorities will provide a reliable and up-to-date source of data for national information systems, including the SISDU system on the IBM 4345. If DGDU can establish a standard format for local information systems, for example, all that will be necessary will be for each city to forward a copy of its current database on disk to DGDU on a regular basis (e.g., every month or every quarter). No work beyond keeping their own database up to date would be required of local planning authorities in order to meet the information needs of DGDU.

Case 20

Computers and development (Panama)

It is probably fair to say that the Panama Canal has been the major factor in the development of Panama up to the present time, and that a continuing supply of fresh water is a prerequisite to the future of the Canal.

The concept of the Canal Area varies somewhat according to usage, but it usually refers to one or more of three overlapping regions — the Canal Zone, the Canal Basin and the Metropolitan Region.

The Canal Zone is the area formerly administered by the Panama Canal Commission. The Canal Zone is the area consisting of the five miles on either side of the Canal from one end to the other plus the associated waterways and shorelines of Lakes Gatun and Alajuela (formerly Lake Madden) and the river joining them (the lower Chagres river). Since the signing of the Torrijos-Carter Treaty in 1979, the Canal Zone has been divided between areas still administered by the Canal Commission (about 520 sq km) and the assets returned or *bienes revertidos* (about 950 sq km) to the government of Panama.

The Canal Basin is a hydrographic region defined by the watershed that feeds the Panama Canal. It covers an area of about 3,300 square kilometres (1,289 square miles), traversing the Canal more or less at right angles and extending from Ciri Grande in the southwest to San Miguel in the northeast. The Basin is divided by the Alajuela Dam into two distinct areas. The area above the dam consists of Lake Alajuela and its tributaries — the upper Chagres, the Pequeni and the Boqueron Rivers. The area below the dam consists of Lake Gatun (through which the Canal itself passes) and its tributaries — the Gatun, Ciri Grande, Trinidad and lower Chagres Rivers. It is the last of these which provides the link through which water flows from the upper to the lower part of the Canal Basin.

Finally, there is the Metropolitan Region or Corridor. This is the planning area generally regarded as consisting of the two cities of Panama and Colon plus the zone of economic activity surrounding and linking them. In physical size and shape, the Metropolitan Region is smaller than the hydrographic Basin; but, while the latter tends to "bulge" in the middle and stops short of the main urban concentrations, the Metropolitan Region is much wider at its northern and southern poles facing the Atlantic and the Pacific Oceans.

However it is defined, though, it would be hard to underestimate the importance to Panama of the Canal Area. By almost any practical standard, the Canal Area is the dominant region of the country. More than half the total population lives within the two provinces (Panama and Colon) that straddle the Canal and almost all of these live within or close to the Canal Area however it is defined. Internal migration flows over the period 1975–80 favoured the Canal Area at the expense of all other regions of the country. Moreover, as is pointed out by the recent USAID-commissioned report on urban development (whence most of the data in this and the next paragraphs are taken),

> Because persons in the prime working age cohorts of 20–39 are the most likely to migrate [the Canal Area] has a larger share of its population in this age bracket than the rest of the country.

By the year 2000, the population of the Canal Area could well equal the 1980 figure for the whole country: that is, about 1.8 million.

Moreover, the Canal Area is estimated to generate more than three-quarters of the total national Gross Domestic Product (GDP). The Canal alone contributes about one-fifth. Estimated growth in real regional product for the period 1978–82 in Panama and Colon was two and four times the average, respectively, in the rest of the country. Mean income per capita in the two provinces is nearly 80 per cent higher than in the rest of the country. In short, it is hard to deny the overwhelming significance to Panama of the Canal Area.

All of this depends in large part on a continuous and substantial supply of fresh water. For every one of the 10–15,000 ships that pass through the Canal each year, a considerable quantity of water (about 400 million litres) is dumped into the sea. Clearly, the Canal is a very large consumer of fresh water in Panama.

The quantity and quality of fresh water in the Canal Area is also critical to the quality of life in numerous other ways. In a direct way, water in the Basin generates some of the electricity that supplies the Canal Area. The Canal Basin is the source of most of the potable water supply for the cities of Panama and Colon. Fresh water is also vital to agriculture, industry, recreation and other activities in the Canal Area. At the broadest level, water also plays a role in determining the climate and carrying capacity of land at a regional and sub-regional scale. Clearly, the quantity and quality of water in the Canal Basin is a matter of critical concern to the whole of Panama.

There appear to be at least two significant threats to the water supply in the Canal Area at this time, both of them caused by deforestation. It is estimated that, since 1900, more than 70 per cent of the rain forest in the Canal Basin has been cut down. Deforestation in the lower (Gatun) basin is almost total (90 per cent); deforestation in the upper (Alajuela) basin is only 20 per cent, but almost all of it has occurred over the past two decades.

The obvious consequence of deforestation is soil erosion and the consequent siltation of the waterways. Although deforestation has been greater in the lower basin, the resulting siltation has been less serious because the terrain is generally flat. Much of the upper basin, on the other hand, is characterized by slopes of 40 and 45 degrees. When the heavy rains typical of the area during the wet season fall on deforested areas, they can very quickly deposit large quantities of sediment in the water courses. According to one study of the upper basin, Lake Alajuela has lost 4.7 per cent of its storage capacity over the past 49 years; most of this must have occurred as a result of deforestation over the past 20 years.

The other important consequence of deforestation, particularly in mountainous areas, is loss of rainfall. Here it is harder to pinpoint causes and effects, but (according to one study cited in the seminar report on *Alternativas de Desarrollo*) rainfall in the Basin has been declining steadily by about 5 mm per year over the past 50 years. It is likely that deforestation is one of the causes.

The main cause of deforestation is the pressure of encroaching human settlements. Thus, the only area remaining largely untouched is the upper Chagres River, the area furthest from areas of settlement. As roads are built into an area, several things happen to threaten the existence of the forests. First, commercial exploitation of the wood becomes economically feasible. Secondly, cattle-raising on pasturelands becomes economically feasible. Thirdly, small-holders (*campesinos*) — many still relying on traditional and destructive techniques of "slash-and-burn" agriculture — can migrate to the area. All of these phenomena have already occurred in the lower (Gatun) part of the Basin; all are even now causing deforestation in the upper (Alajuela) part of the Basin.

Of course, deforestation affects the land as well as the water. Soil erosion inevitably makes land less productive. What is worse is that the returns derived at such cost do not represent the most efficient use of the land, even in the short run. Neither logging, cattle-raising nor slash-and-burn agriculture is as efficient as sound crop management.

Moreover, close to the Canal — and spreading further all the time — deforestation appears to have had the insidious effect of encouraging the spread of a grass known as *hierba mala*, which was originally introduced to help control erosion along the banks of the Canal. This grass (which is related to sugarcane but is not suitable for human or animal consumption) spreads rapidly into open areas, where it soon overwhelms other grasses and bushes and is very difficult to eradicate. What is particularly alarming about the grass is that it makes reforestation very difficult. According to governmental officials, this grass, if allowed to spread unchecked, as it is now, could have the effect of making the region permanently unproductive.

All of this has naturally received a good deal of attention in Panama. There is quite a lot of information about and quite a number of agencies involved in human settlements planning in the Canal Area. Thus, the focus of this report is on the way in which information is used by the Ministerio and some of the other relevant agencies, as they participate in planning and management for the Canal Area.

First of all, the Canal and its water supply naturally dominate all questions of planning and management in Panama. While there is no denying that the Canal and its water are critical factors in the planning of human settlements in Panama, they are not the only ones. Moreover, this is of more than merely academic interest, inasmuch as the way in which problems are perceived tends to affect the way in which information is marshalled and solutions are devised. Consider the following examples:

1. One of the main responses to deforestation has been to focus on the *campesino* as the source of the problem. Thus, various programmes have been implemented (and others contemplated) aimed at persuading and/or compelling him or her to change their way of life. However, such programmes deal with symptoms rather than causes. For the most part, the *campesinos* are not in the Canal doing what they are doing because of ignorance or misunderstanding but because they see no alternative. Indeed, many *campesinos* are capable of providing researchers or television interviewers with quite articulate explanations of what they are doing and why.

   A comprehensive approach to the problem posed by the actions of the *campesinos* in the Canal Area means taking account of factors not peculiar to the Canal Area itself, including "pull" factors (such as urbanization, transportation and access to markets, schools, health services, etc.) and "push" factors (such as land tenure, regional imbalances and poor services and infrastructure in other areas).

2. Similarly, focusing on watershed management leads to an emphasis on safeguarding the water rather than (for example) the

> land. This is not the place to evaluate the various programmes aimed at retarding and reversing the process of deforestation. However, it seems apparent in the ambiguous reaction of some officials to reports of the spread of the *hierba mala* into the upper basin — "We do not regard it as a completely negative development", said one official, since it effectively prevents soil erosion and makes the land unsuitable for destructive agricultural practices — that the broad objectives of development for the Canal Area may have been replaced by a limited set of objectives related specifically to the Canal.

In short, one of the potential obstacles to an effective integration of information and planning for the Canal Area may simply be a conceptual approach that is too narrow. It amounts to the assumption that what is good for the Canal is automatically good for Panama. Nine times out ten, that may be a reasonable assumption; it may also be appropriate for those with special interests in the operation of the Canal. However, it may not be an adequate assumption for human settlements planning in Panama. To put it another way, safeguarding the Canal is always going to be a necessary objective for planning in the Canal Area — but it may not always be a sufficient objective.

The second striking feature of the organization of information about human settlements in the Canal Area is that the number of organizations with a direct interest in it is quite large. Even a partial list of public-sector agencies would have to include those shown in Table 17. Naturally, this list gives no idea of the number and extent of the "informal" groups of interests on the part of land-owners, land-users, industrialists, financiers, labour unions, church groups, student groups, voluntary associations and so on.

In any particular case, of course, the actual number of agencies that will affect or be affected by the outcome of a planning decision affecting the Canal Area will be considerably less than the total, but the important thing is that the interests are fragmented; they do not form a small number of well-organized institutions. The effect of this fragmentation is to make it difficult for information relevant to planning for the area to flow between and among all these governmental agencies and other groups. This is another key obstacle to integration of research and action.

From time to time, there are proposals to reduce some of this fragmentation. For example, Proyecto de Ley No. 21 of 1985 provided for the amalgamation of two key agencies (the Oficina de Planificación y Desarrollo del Area Canalera and the Dirección de Bienes Revertidos) and for a Consejo Consultivo del Area Canalera made up of representatives from three of the Ministerios, the Fuerzas de Defensa

Table 17 Select list of governmental agencies salient to human settlements planning in the Canal Area (Panama)

Ministerio de la Planificación y Politica Economica (including the Oficina de Planificación y Desarrollo del Area Canalera, the Dirección de Asesoria Tecnica Internacional, the Dirección de Planificación y Coordinación Regional, and the Dirección de Programación de Inversiones y Negociación de Prestamos)
Ministerio de Hacienda y Tesorio (including the Dirección de Bienes Revertidos)
Ministerio de Desarrollo Agropecuario (including the Dirección Nacional de Recursos Naturales Renovables and the Dirección de Reforma Agraria)
Ministerio de Vivienda (including the Oficina de Desarrollo Urbano and the Oficina de Investigacciones Urbanas)
Ministerio de Obras Publicas
Ministerio de Educación
Ministerio de Salud
Ministerio de Gobierno y Justicia
Ministerio de Trabajo y Bienestre Social
Panama Canal Commission
Comision Mixta sobre el Medio Ambiente
Comision Nacional sobre el Medio Ambiente
Instituto de Acueductos y Alcantarillados Nacionales
Instituto de Recursos Hidraulicos y Electrificación
Instituto Nacional de Telecommunicaciones
Instituto de Investigación Agropecuaria
Instituto Cartografico Tomy Guardia
Instituto Nacional Cooperativo
Empresa Nacional de Maquinarias Agropecurias
Banco de Desarrollo Agropecuario
Banco Nacional de Panamá
Guardia Nacional/Fuerzas de la Defensa
Universidad de Panamá
Smithsonian Institute Tropical Research Unit
Total = 33

Nacional and the Asamblea Legislativa plus three other persons appointed by the President at his discretion to represent other interests (articles 4 and 17). However, it is clearly impossible (and would probably be undesirable in any case) to rely on legislative action to represent all of the potential interest groups. Thus, if the new law is to play a significant role in co-ordinating governmental agencies and other interests in the Canal Area, the legal provisions will need to be reinforced by operational mechanisms for encouraging the participation of appropriate public and private agencies in the planning and management of the Canal Area.

The third key aspect of an information strategy for the Canal Area is the policy setting. Only a few of the public agencies listed above have published full-scale plans for their activities in the Canal Area. In some cases, no doubt, there are plans, but they are not published. However, in many cases, agencies have either no plans at all or nothing more than guidelines, *normas técnicas* or other equally tentative criteria. The number of real plans of either a regional or sectoral nature actually being implemented in the Canal Area at this time is very limited. Effectively, this was acknowledged in the Proyecto de Ley referred to above, which stipulated that one of the main responsibilities of the new planning agency is to be the preparation of a *Plan General sobre el Uso, Conservación y Desarrollo del Area Canalera* (article 8).

In large part, this lack of planning, compared to research and studies, is just a reflection of the two obstacles already discussed — i.e., lack of an appropriate conceptual framework and excessive institutional fragmentation. But this third problem (lack of a coherent policy setting for development of the area) adds to the difficulties of securing the consensus within the government and in society at large that is necessary for taking action in the Canal Area.

How then can OPDAC help encourage an integration of information and planning in relation to human settlements in the Canal Area? There should be three main priorities for OPDAC:

- a clear definition of overall objectives;
- effective means for co-ordinating salient interests; and
- an overall development strategy.

Several different concepts are competing for attention in the Canal Area. It is important to clarify the scope and objectives of what is being attempted. This means clarifying the geographic scope of the Canal Area, presumably so that it includes both the entire hydrographic basin and the metropolitan region which overlaps it and with which its future is so intimately related. At the same time, it is also going to be important to examine the relationship between the operation of the Canal and broad national objectives — not with a view to sacrificing the interests of the Canal in any sense but rather as a means of determining how much more is involved in national development than preservation of the Canal. The Canal Area is just too closely related to national development in general and human settlements in particular for these processes to be isolated from each other.

It is recommended that steps be taken to create an overall planning agency to co-ordinate governmental actions and to enlist the support and participation of interests in the private and voluntary sectors. It is important that such a co-ordinating agency be that and not more. In other words, the new agency should not be intended to replace or supplant sectoral planning agencies within the government nor to compete with legitimate interests in the private sector. Each of these institutions has its job to do. If anything, the co-ordinating agency should try to assist organizations to improve their planning capacities in relation to the Canal Area, where these capacities may be found to be less than adequate. The co-ordinating agency will clearly need resources and the skills to be able to promote sectoral planning in government and to bring about effective co-ordination among the agencies concerned. This implies a planning agency in which the emphasis is more on planning and management in an organizational sense than on physical design and development in the traditional sense.

It is recommended that an overall development strategy for the Canal Area should be established and should reflect the same kind of preoccupation with the process of implementation as just discussed. Such a plan would provide scope for promoting development of a kind and in areas where it is desirable rather than trying to stop it after it has begun in areas where it is not desirable. The Development Strategy should be a "process plan" rather than a "blueprint plan", concentrating more on the "how" than the "what" and incorporating the development of specific plans as they are needed as part of the process. Thus, the primary purpose of the Development Strategy should be to design and co-ordinate the structures and mechanisms whereby critical resources and critical interests inside government and outside can be brought to bear on the critical problems of the Canal Area. The focus of the Development Plan should be organizations, resources, priorities, timing, etc.; the skills required will be management skills, including administration, finance, community development, public relations, organization development and human relations.

Conclusion

From the foregoing cases, it must be clear that microcomputers have many and varied uses in relation to the management of human settlements. Moreover, it is quite remarkable how effectively microcomputers have, in only a few short years, made what is a relatively "high" technology into something that is entirely appropriate for developing countries. In fact, the key feature of the microcomputer "revolution" may turn out to be not so much its technological as its social achievements. Technologically, microcomputers represent little more than an extension, albeit a dramatic one in terms of size and cost, of what preceded them. From a social standpoint, however, microcomputers have transformed computing from the arcane to the commonplace.

To appreciate the social significance of the microcomputer revolution, consider the following analogy. Imagine a transportation system with its various classes of roads and types of vehicles. Naturally, there are major roads linking large centres of population and a patchwork of minor roads connecting towns, villages and even private houses. Similarly, there are all sorts of vehicles using the roads, including automobiles, buses and trucks. Now, imagine the same system without any minor roads, and without any private cars. All transportation is confined to major roads, and the only vehicles available are buses and trucks. What would that mean?

- First, very few people would have their own personal means of transportation: owning a bus or a truck would be too expensive in relation to the limited use that an individual could make of it.
- Secondly, most driving would be done by specialists; when you wanted to go somewhere, you would have to either use a common carrier travelling along predetermined routes or hire a vehicle and direct the driver to your destination.
- Thirdly, many places would prove difficult or expensive to get to. As long as you wanted to travel between standard destinations

> along standard routes, the system would be efficient (if not always convenient), but, as soon as you wanted to do something out of the ordinary, the cost would increase dramatically.

Clearly, in circumstances like these, a lot of people would go back to walking! What's more, it would probably be the low-income sections of society that would ultimately suffer most from the loss of small-scale, decentralized systems of transport.

Yet, until the early part of this decade, computing was just like this. The only computers were big ones, and applications software was limited to standard packages running standard software. The "personal computer" (unlike the private car) had not even been invented, and the idea of being able to choose between dozens of different roads to get to your destination was completely unknown. Indeed, owing to the fact that computers made by different companies could not even use the same software, rather as if cars made by different companies could not use the same roads, computer use was very limited.

This all changed with the arrival of the microcomputer. Suddenly, computers were affordable: anyone could have one, just as anyone could have a car. Computers were easy to use: ordinary people could reasonably expect to learn to "drive" them without the need to become specialists or technicians. Computers were flexible: users were not limited to a handful of standard applications but could choose among literally thousands of different programmes. To put it simply, microcomputers made computing into a whole new ballgame.

Moreover, these changes were so radical that, as with the invention of the automobile, they are having profound social and cultural effects. Because of the microcomputer, "the practice of computing would become so widespread that the very concept of it would change" and, with it, the user's concept of what he or she is actually capable of (Steven Levy, *Hackers* (New York: Dell, 1984) p. 157, citing Ken Williams, who established On-Line Software, one of the first microcomputer software companies in the USA). Certainly, this was the belief held by some of the pioneers, who

> developed a philosophy about the personal computer and its ability to transform people's lives. [Microcomputers] were amazing not only for what they did, but also for their accessibility . . . people totally ignorant about computers [could learn] to work with them and gain in confidence, so that their whole outlook on life had changed. By manipulating a world inside a computer, people realized that they were capable of making things happen by their own creativity. Once you had that power, you could do anything.
>
> (Ken Williams in Levy, *Hackers*, p. 337)

As another new convert put it, microcomputers seemed to put a "life-loving spirit into the computer".

> . . . if you spent enough time with your [microcomputer], you would realize that *you* could do anything you could think of. [The microcomputer] embodied the essence of pioneering, of doing something brand new, having the courage and the willingness to take risks, doing what's not been done before, trying the impossible and pulling it off with joy. The joy of making things work.
>
> (Margot Tommervik, founder of *Softalk*, in *ibid.*, p. 309)

Thanks to microcomputers, this opportunity was now available to ordinary people. From a medium accessible to only a few, computing had been transformed into one of the mass media, accessible to all. This is a radical change for industrialized countries such as the USA, with all the choices they enjoy; for developing countries, it is a revolution indeed.

Nowhere is the impact of access to computers for the masses likely to be greater than in developing countries, where, because of its expense and complexity, traditional computing has not had much effect on the lives of ordinary people. Microcomputers, however, are neither expensive nor difficult to use. Anyone, anywhere in the world, who can reasonably expect, in terms of cost and ease of use, to have access to an electric typewriter can equally expect to be able to use a microcomputer (unless, of course, his or her government has decided to tax them out of the reach of ordinary people). In many developing countries, particularly in West Asia, India, South-east Asia and Latin America, the use of microcomputers is spreading rapidly in government, business and industry. The incentives are high, not just for organizations in terms of productivity but also for individuals in terms of personal development. For women, young people and others working in traditional bureaucracies, microcomputers offer the chance of rapid advancement. Microcomputers provide a tangible link with an exciting global culture whose underlying message often seems to be, "Look what I can do." Whatever its legality, the availability of "pirated" software in many developing countries, for little more than the price of the disk it is copied on, provides a fertile learning environment, where the user is (at last) on a more or less equal footing with his or her counterpart in the industrialized countries.

Thus, "appropriate" technology does not have to be old-fashioned or unsophisticated. It just has to be cheap, effective, reliable and easy to use. Sometimes that does mean technology invented in medieval times, but appropriate technology can equally mean "space-age" technology such as microcomputers. International development agencies such as

UNCHS have been quick to recognize the significance of microcomputers for development, while developing countries themselves have been eager to gain access to advanced technology.

Of course, microcomputers are not panaceas. They cannot provide for all possible data-processing needs in developing countries, just as private cars cannot replace buses and trucks. However, small personal machines do complement large ones in useful and important ways. In the case of human settlements, there are several factors which help make microcomputers critical to good management:

> First, many human settlements agencies in developing countries are not yet using any kind of computer. So, for them, the first priority is to introduce just the idea of computerization. The easiest and cheapest way to do this is with microcomputers — just as transportation often begins on a small scale.
>
> Secondly, the capacity of microcomputers in the hands of determined users can often be quite astonishing. (In India, a governmental agency decided to computerize all the municipal budgets on an ordinary microcomputer; it took more than 300 floppy disks, but the project succeeded.) A microcomputer is no more a "toy" compared to a big computer than an automobile is a "toy" compared to a bus or a truck.
>
> Thirdly, it is often difficult to anticipate exactly where and how technology will have its biggest impact. The best policy for computers (as for transportation) may well be to create small-scale, decentralized capacity, at least at the outset, so that as many people as possible have access to the technology, and supply can respond to demand with the maximum possible flexibility.
>
> Fourthly, microcomputers have reached the point where their speed and capacity rival those of minicomputers. Many users will find that technology continues to improve quickly enough to keep pace with their capacities and requirements, so that microcomputers are able to meet their needs for years to come.

Even if, in due course, applications are identified for which microcomputers are not adequate, it is almost always helpful to come to such applications by way of experience with microcomputers.

> For one thing, genuine user-needs will probably be clearest and most precise if they are derived through the experience of using a microcomputer-based prototype. It is important to remember that the success of a computer system (like that of a transportation system) depends on a ready supply of inputs (in the case of computers, raw data) and a demand for its output (in the case of computers, useful information). All three items —

input, system and output — are normally easier to design, assess and modify in the context of a microcomputer than of a large machine.

Secondly, experience tends to change the perception of needs. With experience of a particular computer application, users may well find that some capabilities they thought essential when they were working manually are no longer important. Similarly, through experience, users may develop new needs for information which they had not previously imagined could be met. It is often easiest to accommodate the evolution of user-needs by working up through a microcomputer-based prototype system.

Thirdly, with a clear and accurate picture of user-needs, it is easy to identify exactly what additional hardware and software capacity will be required. With mainframe and minicomputers, continuing vendor support and possible incompatibilities with subsequent models are still causes for concern. So, before buying expensive hardware or software, it is desirable to know as clearly as possible what users really need.

In short, even in situations where demand is expected eventually to outgrow the capabilities of microcomputers, it is rash to try to meet those demands without first going through some sort of "learning experience" with microcomputers.

For the management of human settlements, microcomputers offer at least five important advantages when it comes to data-processing. These are improvements in speed, accuracy, security, access and utility.

- First, computerization means speed in processing data. This translates into improved performance of existing operations as well as the possibility of developing entirely new capabilities which would be impractical with manual processing, because of the time and resources that would require.
- Secondly, computerization means accuracy, particularly when transcribing data from one set of records to another. Once data are entered into a computer, they can be transcribed as often as desired and processed in whatever ways may be desired — all without risk of human error. Computers make mistakes but less frequently than is commonly thought. Most "computer errors" are, in fact, due to human errors in data-entry or in software.
- Thirdly, computerization means security of data against physical loss or damage. Manual databases are usually unique, since the cost of maintaining a duplicate is usually high; moreover, the practical need to keep master and duplicate in close proximity to each other inevitably subjects the latter to some of the same risks

as the former (e.g., fire). For all practical purposes, "security through redundancy" is impossible without computerization.

- Fourthly, computerization permits wide access to information about human settlements, both within and outside governmental agencies. Subject to the rights of privacy, up-to-date information on land use (for example) could be made available to all governmental agencies, national and local, concerned with planning as well as to universities, consultants, solicitors, notaries, etc. Again, the practicality of such a proposal depends on computerization.
- Fifthly, computerization permits extensive use of existing data. In the case of a municipal tax system, for example, a computer can help ensure that a tax system is applied fairly and equitably by identifying anomalies in assessed value, land use or sale price. In principle, these calculations can be done by hand, but, in practice, few municipal agencies have the resources to do so without the help of a computer.

What may not be so obvious is the fact that, to some extent at least, all these benefits can be derived from simple microcomputer systems using ordinary, off-the-shelf software, such as database management packages, spreadsheets and geographic information systems. Of course, you can get more benefits from more powerful and more expensive systems, but human settlements agencies can at least start to enjoy the benefits of computerization for no more than the cost of a few electric typewriters.

Microcomputers represent a revolution in information processing, not because of their technology (which is only an extension of what has gone before) but because of their accessibility (which is a quantum leap beyond what has gone before). For example, microcomputer users take their video display screens for granted; they would not dream of running their computers without a screen. Yet, only ten years ago, before the microcomputer revolution began, many computer users worked with only a teletype link to their computer or even without any form of interactive feedback at all. If you made a mistake, you had to wait until you could pick up your printout to discover it! Similarly, almost all microcomputer users know what a spreadsheet programme is, and many use one; yet, before microcomputers were developed, there was no such thing. Spreadsheet programmes had to be adapted from microcomputers to minicomputers and mainframes. In short, it is not computing as such but the use of computing that has been changed so dramatically by the advent of microcomputers.

Moreover, because of their low cost and ease of use, microcomputers represent as much of an opportunity for developing countries as for industrialized ones — perhaps even more of an opportunity for developing countries, since data-processing is less advanced in many of them than it is in industrialized countries. Microcomputers offer developing countries an unprecedented opportunity to capitalize on their often impressive investments in public education. Perhaps microcomputers offer one of those chances that seem all too elusive these days — a chance for developing countries to gain a little ground on the industrial countries, instead of falling further behind.

Bibliography

Part 1: Getting started

Barron, I., and R. Curnow. *The Future with Microelectronics*. London: Pinter, 1979.

Bennet, J.M., and R.E. Kalman, eds. *Computers in Developing Nations*. Amsterdam: North-Holland, 1981.

Berge, Noel, M.D. Ingle and Marcia Hamilton. *Microcomputers in Development: A Manager's Guide*. Rev. ed. Washington DC: Kumerian, 1986.

Bhalla, A., D. James and Y. Stevens. *Blending of New and Traditional Technologies: Case Studies*. World Employment Programme of the International Labour Office. Dublin: Tycooly, 1984.

Cartwright, T.J. "Microcomputers as Appropriate Technology for Human Settlements Planning". *Habitat News* (1985). Pp. 34-5.

Duncan, Ray. *Advanced MS-DOS*. Redmond: Microsoft, 1986.

Friedrichs, G., and A. Schaff, eds. *Microelectronics and Society: for Better or for Worse*. Oxford: Pergamon, 1982.

Gotsch, Carl H. "Applications of Microcomputers in Third World Organizations". *ATAS Bulletin*, 3(June 1986). Published by the UN Centre for Science and Technology for Development. Pp. 69-73.

Harris, Britton. "Does the Third World Need Computers?" Paper prepared for the Joint Meetings of ORSA/TIMS in Atlanta, November 4-6, 1985. Philadelphia: mimeo, 1985.

Levy, Steven. *Hackers: Heroes of the Computer Revolution*. New York: Dell, 1984.

Norton, Peter. *Inside the IBM PC*. Rev. and enlarged. New York: Prentice-Hall, 1986.

——. *PC-DOS: Introduction to High-Performance Computing*. New York: Prentice-Hall, 1985.

——. *Programmer's Guide to the IBM PC*. Bellevue: Microsoft, 1985.

Schware, Robert, and Alice Trembour. "Rethinking Microcomputer Technology Transfer to Third World Countries". *Science & Public Policy*, February 1985. Pp. 15-20.

Turkle, Sherry. *The Second Self: Computers and the Human Spirit*. New York: Simon and Schuster, 1984.

Wolverton, Van. *Running MS-DOS*. Bellevue: Microsoft, 1984.

Case 1: Preparing for a consultant (Jamaica)

Ministry of Finance and Planning. Town and Country Planning Department. *Annual Report*. Kingston: mimeo, 1984.

Case 2: Now or later? (Trinidad and Tobago)

Oberlander, H. Peter. *A Report on the Establishment and Organization for Planning for Urban and Regional Development in Trinidad and Tobago*. Report for the United Nations Programme for Technical Assistance, Department of Economic and Social Affairs. New York: mimeo, 1962. 56 pp.

Statskonsult International AB. *The Administrative Improvement Program*. 10 vols. Trinidad: mimeo, January 1980. Approx. 800 pp.

Trinidad and Tobago, Republic of. *Commission of Enquiry into Land and Building Use*. Joseph Crooks, Chairman. Interim Report. Trinidad: Government Printery, 1981. 93 pp.

——. *The Imperatives of Adjustment: Draft Development Plan, 1983-1986*. Report of the Task Force Appointed by Cabinet to Formulate a Multi-Sectoral Development Plan for the Republic of Trinidad and Tobago. William G. Demas, Chairman. Two vols. Trinidad: Government Printery, 1984. 235 pp.

——. *Report of the Cabinet-Appointed Committee to Prepare a Policy Paper on Requests for the Construction of Roads on Private Lands*. I. Mohammed, Chairman. Trinidad: mimeo, 1983. 34 pp.

——. Ministry of Finance and Planning. Town and Country Planning Division. *National Physical Development Plan Trinidad and Tobago: Planning for Development*. Vol. 1 — Survey and Analysis. Vol. 2 — Strategies and Proposals. Trinidad: Superservice, September 1982. 272 pp.

——. *Planning for Development: the Capital Region*. Draft Local Plan No. T1/2/3. Trinidad: Superservice, September 1975. 147 pp.

——. *Planning for Development: the Caroni Region*. Draft Local Plan No. T5. Trinidad: Superservice, June 1974. 150 pp.

——. *Report on Development Planning — the Town and Country Planning Division, 1974-79 — for the Commission of Enquiry into Land and Building Use*. Internal report. Trinidad: mimeo, December 1979. 18 pp.

Case 3: Assessing the plan (Tunisia)

Centre national de l'informatique. Projet d'automatisation de la Direction de l'aménagement du territoire. *Convention d'assistance à la réalisation d'un système informatique*. Tunis: April 1983. 17 pp.

——. *Etude de l'existant*. Tunis: September 1983. 52 pp. plus Annex.

——. *Sous-système de suivi des plans d'aménagement: Conception et étude détaillée d'un nouveau système*. Tunis: January 1984. 179 pp. plus Annex.

——. *Sous-système de representation graphique: Conception et étude détaillée d'un nouveau système*. Tunis: June 1984. 104 pp. plus Annex.

——. *Plan informatique 1984–88*. Tunis: March 1984. 100 pp.

——. *Cahiers de charge pour l'acquisition de matériel informatique.* Tunis: April 1985. 60 pp.
Zaiem, M.H. *Construction d'un indicateur synthétique de "développement régional" par analyse en composantes principales.* Report to the Commissariat-général du développement régional. Tunis: no date. 14 pp.

Case 4: Taking delivery (Abu Dhabi)

Abu Dhabi Industrial Development Corporation (ADIDCO). "Supply of a Computer System for the Department of Public Works, Abu Dhabi". Abu Dhabi: September 1984.
EFHA Consulting Engineers. "Specifications for a Computer System for the Department of Public Works, Abu Dhabi". Abu Dhabi: January 1984.

Part 2: Computer applications

Brooner, E.G. *Microcomputer Data-Base Management.* Indianapolis: Howard Sams, 1982.
Byers, Robert A. *Everyman's Database Primer, Featuring dBaseII.* Reston: Reston Publishing, 1983.
Carpenter, James, Dennis Deloria and David Morganstein. "Statistical Software for Microcomputers". *BYTE* (April 1984). Pp. 234–64.
Castro, Luis, Jay Hanson and Tom Rettig. *Advanced Programmer's Guide Featuring dBaseIII and dBaseII.* Culver City: Ashton-Tate, 1985.
Cobb, Douglas, with S.S. Cobb and Gena Cobb. *Douglas Cobb's 1-2-3 Handbook.* New York: Bantam, 1986.
Danish Hydraulics Institute. *MOUSE: Microcomputer Urban Sewer Program.* Computer program and manual. Version 1.0 for PC/MS-DOS. Copenhagen: DHS, 1986.
Date, C.J. *An Introduction to Database Systems.* Reading: Addison-Wesley, 1987.
Freiling, Michael J. *Understanding Data Base Management.* Sherman Oaks: Alfred Publishing, 1982.
Griesmer, J.R. *Microcomputers in Local Government.* Washington DC: International City Management Association, 1983.
Groff, W. De. "Microcomputers and Their Impact on Local Government Operations". *Computers, Environment and Urban Systems* 8(1983). Pp. 45-9.
Henderson, T.B., D.F. Cobb and G.B. Cobb. *Spreadsheet Software: From Visicalc to 1-2-3.* Indianapolis: Que Corporation, 1983.
International City Management Association. *Software Reference Guide 1986/87.* Dennis M. Kouba, ed. Philadelphia: ICMA, 1986.
Kammeier, H. Detlef. "Microcomputer-Based Assessment of the Accessibility of Central Place Services". *The Asian Geographer* (September 1986).
Kelly, Susan Baake. *Mastering Word Perfect.* Alameda: Sybex, 1986.
Kruglinski, David. *Data Base Management Systems: a Guide to Microcomputer Software.* Berkeley: Osborne/McGraw-Hill, 1983.

Kruijff, G.J.W. de. "Urban Cost Model Program". Microcomputer program based on the Bertaud Model, with manual. Version 1.0. Jakarta: mimeo and disk, October 1984. Manual 24 pp.
Lachenbruch, P.A. "Statistical Programs for Micro-Computers". *BYTE* (November 1983). Pp. 560–70.
Landis, John D. "Electronic Spreadsheets in Planning: the Case of Shift-share Analysis". *Journal of the American Planning Association*, 51.2(Spring 1985). Pp. 216-24.
Levine, Harvey A. *Project Management Using Microcomputers*. Berkeley: Osborne/McGraw-Hill, 1986.
Levine, Ned. "The Construction of a Population Analysis Program Using a Microcomputer Spreadsheet". *Journal of the American Planning Association*, 51.4(Autumn 1985). Pp. 496-511.
Martin, James. *Managing the Data Base Environment*. Englewood Cliffs: Prentice-Hall, 1983.
"Microcomputer Software Packages Produced by the Organizations of the United Nations System". Supplement to *ACCIS Newsletter* 4.3 (September 1986).
Naiman, Arthur. *Mastering Wordstar on the IBM PC*. 2nd ed. Alameda: Sybex, 1987.
Newton, P.W., and M.A.P. Taylor, eds. *Microcomputers for Local Government Planning and Management*. Melbourne: Hargreen, 1986.
Ontario Ministry of Municipal Affairs and Housing. *Microcomputers in Small Municipalities: a Guide*. Toronto, 1984.
——. *The Use of Computers in Planning*. Toronto: 1983.
Ottensmann, John R. "Analyzing Planning Alternatives Using Electronic Spreadsheets". *Journal of Planning Education and Research*, 4.1 (August 1984). Pp. 33-42.
——. *BASIC Microcomputer Programs for Urban Analysis and Planning*. New York: Chapman and Hall, 1985.
——. "Microcomputers in Applied Settings: the Example of Urban Planning". *Sociological Methods in Research* 9(1983). Pp. 493–501.
——. *Using Personal Computers in Public Agencies*. New York: Wiley, 1985.
Poole, Lon. *Using Your IBM Personal Computer*. Indianapolis: Howard Sams, 1983.
Rinearson, Peter. *Word Processing Power with Microsoft Word*. Redmond: Microsoft, 1986.
Sawicki, David S. "Microcomputer Applications in Planning". *Journal of the American Planning Association*, 51.2(Spring 1985). Pp. 209–15.
Schneider, David I. *Handbook of BASIC for the IBM PC*. New York: Prentice-Hall, 1985.
Simpson, Alan. *Understanding dBASE III Plus*. Alameda: Sybex, 1986.
——. *Understanding R:base System V*. Alameda: Sybex, 1987.
Sipe, Neil, and Robert W. Hopkins. *Microcomputers and Economic Analysis: Spreadsheet Templates for Local Government*. Bureau of Economic and Business Research Monograph No. 2. Miami: University of Florida, 1984. 152 pp.

Stultz, Russell A. *The Illustrated dBaseII Book*. Plano: Wordware, 1984.
United Nations Centre for Human Settlements (Habitat). *HFSL: Housing Finance Savings and Loans*. Computer program and manual. Version 2.0 for PC/MS-DOS. Nairobi: 1987.
——. *UDMS — Urban Data Management System*. Computer program and manual. Version 5.3 for PC/MS-DOS. Nairobi: 1987.
United Nations Development Programme and The World Bank. *Urban Water Supply and Sewage System Modeling Program*. Computer program and manual. New York: 1986.
United States. Agency for International Development. *Preparing a National Housing Needs Assessment*. Occasional Paper Series. Prepared by Robert R. Nathan Associates Inc. and The Urban Institute. Washington DC: mimeo, March 1984. 85 pp.
——. Department of Transport. *Quick Response System (QRS) for Travel Demand Analysis*. User's Manual. Washington DC: Federal Highway Administration, 1985.
Williams, R.E., and B.J. Taylor. *The Power of Supercalc*. Portland: Management Information Source, 1982.
Wood, Chris. *The SuperCalc Program Made Easy*. 2nd. ed. Berkeley: Osborne/McGraw-Hill, 1987.
World Bank. *The Urban Edge*. (Special Issue on the Bertaud Model.) 10.1(January 1986). 12 pp.
——. and PADCO Inc. *The Bertaud Model: A Model for the Analysis of Alternatives for Low-Income Shelter in the Developing World*. Urban Development Department Technical Paper No. 2. Washington DC: World Bank, December 1981. 147 pp.
——. *The Bertaud Model, including Affordability and Site Layout Submodels*. Computer program and manual. Washington DC: 1983.

Case 5: An embarrassment of choices (Burma)

Asian Development Bank. *Economic Survey on Burma*. Report BUR-Ec-2s-3. Manila: mimeo, June 1984. 48 pp.
Lemarchands, Guy. "Evaluation of Project BUR/80/005: Rangoon City and Regional Development Phase I and Reformulation of Project BUR/85/016: Rangoon City and Regional Development Phase II". For the United Nations Development Programme. Rangoon: mimeo, April 1986. 37 pp.
Ministry of Construction. Housing Department. [Briefing Notes from the Various Divisions prepared for the UNCHS (Habitat) Data Management Advisory Mission.] Rangoon: manuscript, 1986.
——. *Draft Report on Information System for Urban and Regional Planning*. Urban and Regional Planning Division. Rangoon: mimeo, October 1986. 12 pp.
——. *Map Information Recording System*. Urban and Regional Planning Division. Rangoon: mimeo, October 1986. 20 pp.
——. *Rangoon City Structure Plan*. Volume 1. Report prepared under UNDP/UNCHS Project BUR/80/005. Rangoon: mimeo, June 1986. 138 pp.

——. *The Housing Department: a Preliminary Evaluation.* Report prepared under UNDP/UNCHS Project BUR/80/005. Rangoon: mimeo, March 1984. 60 pp.

Thein, U Kyaw. "Country Background Paper [on Burma]". Paper for the Regional Seminar on Financing of Low-Income Housing, Manila, February 7-12, 1983. Rangoon: mimeo, no date.

United Nations Development Programme. Reference and Documentation Unit. "Explanation on Fifth Four-Year Plan Targets" and other reports on the discussion of the Report of the Council of Ministers to the Fourth Pyithu Hluttaw. Translated from the *Guardian* of March 11, 14 and 15, 1986. Rangoon: mimeo, 1986. 7 pp.

World Bank. *Burma: Policies and Prospects for Economic Adjustment and Growth.* Report No. 4814-BA. Dated November 18, 1985. Washington DC: World Bank, 1985. 137 pp.

Case 6: Physical planning applications (Yemen AR)

Ministry of Municipalities and Housing. Hodeidah Urban Development Project. *Third Six-Monthly Progress Report for the Period Ending 31st December 1985.* Sana'a: January 18, 1986. 21 pp.

——. Sana'a Urban Development Project. *Eighth Six-Monthly Progress Report for the Period Ending 31st December 1985.* Sana'a: January 18, 1986. 18 pp.

Switzerland. Directorate of Development Cooperation and Humanitarian Aid (DCA/DEH). *An Analysis of a Suitable Land Registration System for, and its Implementation in, the Yemen Arab Republic.* Report to the Survey Authority of the Ministry of Public Works based on a Mission by David Hughes, July 20-September 14, 1983. Bern: mimeo, 1983. 85 pp. and appendices.

World Bank. *Yemen Arab Republic. Sana'a Urban Development Project.* Staff Appraisal Report. Report No. 3642-YAR. Washington DC: December 30, 1981. 60 pp.

——. *Yemen Arab Republic. Second Urban Development Project.* Staff Appraisal Report. Report No. 4675-YAR. Washington DC: November 9, 1983. 47 pp.

Case 7: Social planning applications (Malaysia)

Ministry of Housing and Local Government. [UNCHS Mission Briefing Notes.] Prepared by the Computer Unit of the Research Division. Kuala Lumpur: photocopy, no date. 4 pp.

Town and Country Planning Department. *Annual Report 1985.* Kuala Lumpur: photocopy, no date. 14 pp.

——. *Briefing Note on the Federal Town and Country Planning Department.* Kuala Lumpur: photocopy, June 1986. 14 pp.

——. *Micro Computers for the Town and Country Planning Department.* Kuala Lumpur: photocopy, no date. 5 pp.

Case 8: Management applications (Thailand)

Ministry of Communications. Department of Land Transport. [Untitled Summary of Departmental Responsibilities and Structure.] Bangkok: mimeo, August 1984 [BE 2527]. 8 pp.

——. Land Transport Act, 1979 [BE 2522]. Translated into English from the Government Gazette of March 21, 1979. Volume 96 (Special Issue), part 38. Bangkok: mimeo, no date. 34 pp.

——. [Statistical Report of the Changwat of Lamphun for the Year 1984 (BE 2527)]. In Thai. Lamphun: mimeo, 1984. 76 pp.

Part 3: Information systems

American Farmland Trust. *Survey of Geographic Information Systems for Natural Resources Decision Making at the Local Level*. Washington DC: mimeo, August 1985.

Cartwright, T.J. "Information Systems for Planning in Developing Countries: Some Lessons from the Experience of the United Nations Centre for Human Settlements". *Habitat International* 11.1(1987).

——. "Microcomputers as Appropriate Technology for Human Settlements Planning". *Habitat News* (1985).

Cowen, D.J., A H. Vang and J.M. Waddell. "Beyond Hardware and Software: Implementing a State-Level Geographic Information System". In E. Teicholz and B.J.L. Berry, eds. *Computer Graphics and Environmental Planning*. Englewood Cliffs: Prentice-Hall, 1983.

Dutton, W.H. "Eighty Thousand Information Systems: the Utilization of Computing in American Local Government". *Computers, Environment and Urban Systems* 7(1982). Pp. 21–33.

Kraemer, K.L., ed. "Municipal Information Systems". Special issue of *Computers, Environment and Urban Systems* 7(1982). Pp 1–140.

Minieka, Edward. *Optimization Algorithms for Networks and Graphs*. New York: Marcel Dekker, 1978.

Oppenheim, N. *Applied Models in Urban and Regional Analysis*. Englewood Cliffs: Prentice-Hall, 1980.

Ottensmann, John R. *BASIC Microcomputer Programs for Urban Analysis and Planning*. New York: Chapman and Hall, 1985.

Rushton, G. *Optimal Location of Facilities*. Wentworth: COMPress, 1979.

United Nations Centre for Human Settlements (Habitat). "A Microcomputer Physical Planning System from Habitat". *Technical Note No. 10*. Nairobi: UNCHS, 1988.

United Nations Educational Scientific and Cultural Organization (UNESCO). *Conceptual Framework and Guidelines for Establishing Geographic Information Systems*. Prelim. version. Ed. by W.H. Erik de Man. Paris: UNESCO, August 1984.

Wegener, M. "Urban Information Systems and Society". In J.F. Botchie, P. W. Newton, P. Hall and P. Nijkamp, eds. *The Future of Urban Form: the Impact of New Technology*. London: Croom Helm, 1985.

Wiggins, Lyna L. "Three Low-Cost Mapping Packages for Microcomputers". *APA Journal* (Autumn 1986). Pp. 480–8.

Case 9: Data aren't everything (Indonesia)

Automated Project Management System Study. "Computer Mode Study" and "Automation Strategy for Cipta Karya's Project Management System". By SGV and Co. Two vols. June and November 1984. 24 and 18 pp.

National Urban Development Strategy (NUDS). "Big Towns, Intermediate Towns, Small Towns — No Towns: Some Preliminary Thoughts on the Urban Hierarchy in Indonesia and Its Implications for Future Planning". By Terry G. McGee. No date (1984?). 18 pp.

——. "The National Urban Development Strategy Project: an Overview". Project Staff Report. June 2, 1983. 7 pp.

——. "Outline Urban Strategy". Draft Main Report. October 1983. 144 pp.

——. "Preliminary Computer System Management Plan". March 5, 1984. 6 pp.

——. "Resource Requirements for Information System Development". Memo from G.T. Kingsley to T. Akita. May 12, 1984. 3 pp.

——. "Transport Patterns and Trends" and "Overview of Current Transport Infrastructure Capacity". By Hasfarm Dian in association with Hoff and Overgaard. Two working papers. October-November 1984. 115 and 105 pp.

——. "Urban Characteristics and Typologies in Indonesia". By Institut Teknologi Bandung. Main Report (Draft) of the Urban Systems and Structures Subcontract. No date (1985?). Approx. 255 pp.

——. "Urban Services Program Accomplishments and Trends". By PADCO/P.T. Dacrea. Final Report. Four vols. August-December 1984. 200, 38, 250 and 177 pp. approx.

——. "Urban Strategies and Integrated Urban Development". By G.T. Kingsley. Paper presented to the Workshop on Integrated Urban Development Programme Approach in Urban Services (Bogor, July 23-4, 1984). 11 pp.

——. "Urbanization and Structural Change in Employment in Indonesia: Evidence from Provincial and Kabupaten-Level Analysis of 1971 and 1980 Census Data". By Gavin W. Jones *et al.* November 1984. 96 pp.

United Nations Centre for Human Settlements (Habitat). "Urban Development in Indonesia: an Approach to Project Formulation Based on Outputs from the NUDS". By Kenneth Watts. Report on a Mission from March 9-28, 1984. 15 pp.

——. "Mission Report: NUDS Preliminary Computer Plan, Jakarta". By Philippe Billot. Report on a Mission from August 11 to September 1, 1984. 36 pp.

United Nations Development Programme. "Development of the Government Information Network System". Project Document: Project INS/82/007. September 1983. 16 pp.

World Bank. "Urban Information System Feasibility Study: Draft Terms of Reference". No date (1985?). 8 pp.

Case 10: Top-down versus bottom-up design (India)

British Council. "Madras Metropolitan Development Authority — Report on a Visit to India". By E.M. Davies. Report No. T2471 for the British Council. London, June 1984. 76 pp.
Gwin, Catherine, and L.A. Veit. "The Indian Miracle". *Foreign Affairs* (Spring 1985).
Gujarat Town Planning and Valuation Department. "Pilot Study on Urban Information System: Case Study of Anand — Agency Analysis Report". Ahmedabad, August 1984. 198 pp.
——. "Pilot Study on Urban Information System: a Case Study of Anand — System Requirements". Draft Report. Ahmedabad, 1984. 52 pp.
Institute for Coastal and Offshore Research. *Remote Sensing for Planning and Environmental Aspects of Urban and Rural Settlements.* Technical Papers for a Conference in January 1985. Two vols. Visakhapatnam, January 1985. Approx. 800 pp.
National Informatics Centre. *Status Report.* New Delhi: INSDOC, May 1983. 119 pp.
National Institute on Urban Affairs. *Municipal Finance in India: Sources of Revenue and Components of Expenditure, 1979–80.* Report to the Eighth Financial Commission. Four vols. New Delhi, May 1984. Approx. 1,200 pp.
Tamil Nadu Directorate of Town and Country Planning. "Pilot Study on Urban Information System: a Case Study for Chengalpattu — Agency Analysis Report". Madras, December 1982. 248 pp.
——. "Pilot Study on Urban Information System: a Case Study of Chengalpattu. Report on Attribute Analysis and Format Design". Two vols. Madras, February 1984. 39 and 158 pp.
Town and Country Planning Organization. "Design, Development and Implementation of Urban Information Systems (URBIS)". Notes of a Workshop at Visakhapatnam held in May 1983. New Delhi, May 1983.
——. "Proposal for Purchasing an Appropriate Computer System for TCPO, New Delhi". Internal discussion paper. New Delhi, 1984(?). 4 pp.
——. Urban and Regional Information System. "Report of the Steering Group". New Delhi, April 1981.
——. "Report of the Working Group on a Regional Information System". Draft. New Delhi, March 1983.
United Nations Centre for Housing, Building and Planning. *Feasibility of Information Systems to Support Planning Functions in India.* By Jerry Coiner. Report on a Mission. New York, March 1977. 40 pp.
World Bank. "Situation and Prospects of the Indian Economy — a Medium-Term Perspective". Three vols. Report No. 4962-IN. Washington DC, April 1984. 28 and 244 pp.

Case 11: Data banks and information systems (Philippines)

Ministry of Human Settlements. *Annual Report 1983.* For the period June 1978 to December 1983. Manila: University Publishing House, 1984. 50 pp.

——. *National Multi-Year Human Settlements Plan, 1983–1987 and 2000: a Physical Development Framework.* Manila: mimeo, 1983. 259 pp. Accompanied by thirteen regional volumes.

——. Information Systems Technical Services. *User's Manual for the Numerical Data Base in the Planner's Data Bank.* Manila: mimeo, November 1978. 86 pp.

——. Community Organization and Development Group. Program Management Division. *Services Delivery Profile.* Four vols. Manila: typescript, October 1984.

——. Strategic Planning Group. *Computer Data Bases and Software Bank.* Manila: mimeo, September 1983. 14 pp.

——. *MHS Data Banking Projects.* Set of documents pertaining to six ongoing data-banking projects. Manila: mimeo, 1985.

——. *An Overview on the Use of Computers for Planning at the Ministry of Human Settlements.* Manila: mimeo, no date (May 1985?). 25 pp.

Raralio, Encarnacion N., ed. *Philippine Shelter System and Human Settlements.* Human Settlements Monograph Series No. 1/1983. Manila: Rapid Publishing, October 1983. 278 pp.

World Bank (IBRD). *The Philippines: an Agenda for Adjustment and Growth.* Report No. 5258-PH. Washington DC: mimeo, November 1984. 170 pp.

——. *Philippines Municipal Development Project: Staff Appraisal Report.* Report No. 5027-PH. Washington DC: mimeo, May 1984. 71 pp.

——. *Regional Cities Development Project.* Report No. 4094-PH. Washington DC: mimeo, March 1983.

Case 12: A land management information system (Mauritius)

Mauritius, Colony of (?) *The Domaine Book.* Inventory of Crown Lands in Mauritius as of 1934. Port Louis(?): no publisher, no date (*ca.* 1935).

Meade, J.E., *et al. The Economic and Social Structure of Mauritius.* Report to the Governor of Mauritius. London: Frank Cass, 1961. 247 pp.

Ministry of Economic Planning. *1984–1986 Development Plan.* Port Louis: Government Printer, January 1985. 202 pp.

Ministry of Housing and Lands. *Report on Matters concerning the Registration of Title to Land and the Cadastral Map of Mauritius.* By L.S. Himely. Port Louis (?): no publisher, *ca.* May 1961. 65 pp.

Ministry of Housing, Lands and the Environment. *Report of the Fact-Finding Committee on Crown Lands.* Jairaj Ramkissoon, Chairman. Two vols. Port Louis: mimeo, June 1985. 360 and 353 pp.

Patten, Nandraj. "An Overview of Physical Planning in Mauritius". Paper for Physical Planning Development Seminar, 1–5 April 1985, sponsored by the Ministry of Housing, Lands and the Environment. Port Louis: mimeo, 1985. 25 pp.

——. "Training Programme Followed by Nandraj Patten, Town and Country Planning Officer, Ministry of Housing, Lands and the Environment, July 8-October 4, 1985". Report to UNCHS/UNDP Project on Development Planning Training Programme (MAR/83/001). Port Louis: mimeo, October 1985. 26 pp.

Research Triangle Institute. "Analysis of Requirements for Computerized Information Management in the Mauritius Housing Corporation and the Ministry of Housing, Lands and the Environment". By Jerry Van Sant. Research Triangle, NC: typescript, March 1984. 28 pp.
Titmuss, Richard M., and Brian Abel-Smith. *Social Policies and Population Growth in Mauritius*. Report to the Governor of Mauritius. Sessional Paper No. 6 of 1960. London: Frank Cass, 1961. 308 pp.

Part 4: Institutional factors

Durr, Michael. *Networking IBM PCs*. 2nd. ed. Indianapolis: Que, 1987.
Glossbrenner, Alfred. *The Complete Handbook of Personal Computer Communications*. New York: St Martin's, 1985.
Gofton, Peter W. *Mastering Crosstalk XVI*. Alameda: Sybex, 1987.
Sime, M.S., ed. *Designing for Human-Computer Communication*. London: Academic Press, 1983.

Case 13: Identifying user needs (Bahrain)

Council of Ministers. Central Statistics Organization. *Statistical Abstract, 1983*. Manama: Government Press, December 1984. 354 pp.
Gredeco-Ansari Consultants. *Rural Rehabilitation Programme*. Report to the Ministry of Housing. Manama: mimeo, August 1982. 54 pp.
Linn, Johannes F. *Cities in the Developing World: Policies for Their Equitable and Efficient Growth*. World Bank Research Publication. New York: Oxford, 1983.
Ministry of Cabinet Affairs. Directorate of Statistics. *Bahrain Census of Population and Housing — 1981*. Manama: Arabian, 1982. 510 pp.
Ministry of Housing. *A National Land Use Plan for Bahrain*. Draft for Discussion. Manama: mimeo, no date. 17 pp.
——. *An Urban Renewal Strategy for Muharraq, 1984–1999*. Manama: mimeo, no date. 88 pp.
Ministry of Works, Power and Water. Central Planning Unit. *The Allocation, Acquisition and Registration of Land for Public Utilities*. A Working Guide for Staff. Manama: mimeo, October 1985.
——. National Strategy for Sewerage and Sewage Treatment. *Preliminary Block by Block Population Projections*. Manama: mimeo, December 1984. 31 pp.
Reutlinger, Shlomo. *Techniques for Project Appraisal under Uncertainty*. World Bank Staff Paper No. 10. Baltimore: Hopkins, 1970. 95 pp.
United Nations Centre for Human Settlements (Habitat). *National Land Use Plan*. Section I: Generalized Land Use Changes, 1971–1982. By T.G. Alexander. Report for the Ministry of Housing. Manama: mimeo, June 1983. 39 pp.
——. *National Land Use and Physical Development Plan. Possible Distributions of Population and Employment to 1991; Monitoring of Residential and Non-Residential Development*. By J.A.S. Tennert.

Working Paper III for the Ministry of Housing. Manama: mimeo, July 1982. 205 pp.
——. *Projection of Population Composition and Distribution, State of Bahrain, 1986 to 2001. Principal Directions, Pitfalls and Final Findings.* By A.M. Farrag. 2 vols. Report for the Ministry of Housing. Manama: mimeo, October 1982. 89 pp.
World Bank. *Bahrain Current Economic Position and Prospects.* Report No. 2058-BH. Washington DC: IBRD, June 1978. 54 pp. and appendices.

Case 14: Setting priorities (Turkey)

Celasun, Merih. *Sources of Industrial Growth and Structural Change: the Case of Turkey.* World Bank Staff Working Paper No. 614. Washington DC: IBRD, 1983. 169 pp.
Kentkur a.s. *Kentkur New Bulletin.* Special Issue on Kentkoop. Ankara: Rekmay, May 1986. 32 pp.
Ministry of Public Works and Settlement. *Activities for the International Year of Shelter for the Homeless in Turkey.* Position Paper for the UNCHS 9th Session, Istanbul. Ankara: mimeo, 1986. 18 pp.
——. *Organization and Functions [of the Ministry].* Background Report for the UNCHS 9th Session, Istanbul. Ankara: Barok, 1986. 32 pp.
Turkey, Government of. *Fifth Five-Year Development Plan, 1985–1989.* Ankara: Basimevi, 1985. 224 pp.
United States. Department of Transport. *Quick Response System (QRS) for Travel Demand Analysis. User's Manual.* Washington DC: Federal Highway Administration, 1985.
World Bank. *Turkey — the Vth Five-Year Plan in the Context of Structural Adjustment: a Review.* Report No. 5418-TU. Three vols. Washington DC: IBRD, July 20, 1985. 21, 216 and 112 pp.

Case 15: Increasing productivity (Swaziland)

Ministry of Finance and Planning. *Fourth National Development Plan, 1983/4-1987/8.* Mbabane: Government Publisher, 1983.
Ministry of Natural Resources, Land Utilization and Energy. *National Physical Development Plan.* Draft version. Mbabane: typescript, April 1986.
United Nations Centre for Human Settlements (UNCHS). "Introduction of Microcomputer-Based Planning Information Systems". By Jerry C. Coiner. Report of a Special Advisory Mission to the Ministry of Natural Resources, Land Utilization and Energy. Mbabane: June 1982.

Case 16: Improving computer use (Jordan)

Housing Corporation. *The Role of the Housing Corporation in the Housing Sector.* By Ahmad Al Fandi. Amman: mimeo, 1981. 115 pp.
——. [Annual Report of the Housing Corporation, 1983]. Draft manuscript. Amman: typescript, 1984.

Johansson, Jan-Henrik. "Simultaneous Equations with Lotus 1-2-3". *BYTE* (February 1985). Pp. 399–405.
National Planning Council. *Five-Year Plan for Economic and Social Development, 1981–1985*. Amman: Royal Scientific Society Press, no date. 372 pp.

Case 17: Building institutional capacity (Sri Lanka)

Jayaweera, D.S. "Identification of an Appropriate Urban Data Base for Effective Urban Planning". M.Sc. Dissertation submitted to the University of Moratuwa. Colombo, December 1984. 173 pp.
Leitan, G.R. Tressie. *Local Government and Decentralized Administration in Sri Lanka*. Colombo: Lake House, 1979.
Mendis, M.W.J.G. *Local Government in Sri Lanka*. Colombo: Apothecaries, 1976.
Ministry of Local Government, Housing and Construction. *Performance Improvement of Urban Local Authorities: the Draft Project Report*. Colombo, no date. 22 pp.
Sri Lanka, Government of. *Urban Development Authority Law, No. 41 of 1978 (Incorporating Amendments made by Act No. 70 of 10979 and Act No. 4 of 1982)*. Colombo, September 1983. 31 pp.
United Nations. Department of Technical Cooperation for Development. *Mission Report — Colombo, Sri Lanka*. By Peter Browne. Report by the Interregional Adviser on Information Management and Use of Computers in Public Administration. New York, December 1984. 35 pp.
World Bank. *Sri Lanka Urban Sector Report*. Report No. 4640-CE. Including Draft Tables for a Management Information System for Urban Local Authorities (Annex 2). Washington DC, June 1984. 103 pp.

Part 5: Policy choices

Cartwright, T.J. "The Lost Art of Planning". *Long Range Planning* 20.2(1987). Pp. 92–9.
Cordell, Arthur J. *The Uneasy Eighties: the Transition to an Information Society*. Science Council of Canada Background Study No. 53. Ottawa: Supply and Services Canada, 1985.
Forester, T., ed. *The Microelectronics Revolution*. Oxford: Blackwell, 1980.
International Telecommunications Union. *The Missing Link*. Geneva: ITU, 1986.
Kaplan, Abraham. *The Conduct of Inquiry*. San Francisco: Chandler, 1964.
Langendorf, Richard. "Computers and Decision Making". *Journal of the American Planning Association* 51.4(Autumn 1985). Pp. 422–32.
Levy, Steven. *Hackers: Heroes of the Computer Revolution*. New York: Dell, 1984. 448 pp.
Sawicki, David S., and William R. Page. "Teaching Computer and Policy Analysis Skills in a Case Study Course". *Journal of Planning Education and Research* 4(August 1984). Pp. 43–54.

Toffler, Alvin. *The Third Wave*. New York: Random House, 1980.
Turkle, Sherry. *The Second Self: Computers and the Human Spirit*. New York: Simon and Schuster, 1984.

Case 18: Maintaining the balance of power (Cyprus)

LK Computer Systems Ltd. "Department of Town Planning and Housing Planning Application System Specification". Nicosia: mimeo, February 1983. 20 pp.
United Nations Development Programme and United Nations Centre for Human Settlements (Habitat). *Nicosia Master Plan*. Final Report. Two vols. Nicosia: mimeo, July 1984. 198 and 35 pp.
——. *Follow-Up Planning Activities* and *Work Programme for Activities of Common Interest*. Nicosia: mimeo, October and November 1985. 8 and 5 pp.

Case 19: Computers and decentralization (Mexico)

Banco Interamericano de Desarrollo. "Mexico". In *Progreso Económico y Social en América Latina: Integración Económica — Informe 1984*. Washington DC: BID, 1984. Pp. 355–62.
Banco de Mexico. *Informe Anual, 1984*. Mexico City: 1985. 255 pp.
Comisión Económica para América Latina y el Caribe (CEPAL). *Notas para el Estudio Economico de America Latina y el Caribe, 1984: Mexico*. No. LC/MEX/L.5. Mexico City: April 25, 1985. 67 pp.
International Bank for Reconstruction and Development (World Bank). *Mexico: Future Directions of Industrial Strategy*. No. 4313a-ME. Washington DC: May 31, 1983. 121 pp.
Lee, Colin. "Gravity Models". In *Models in Planning: an Introduction to the Use of Quantitative Models in Planning*. Urban and Regional Planning Series, Vol. 4. Oxford: Headington, 1977. Pp. 57–88.
Mexicanos, Estados Unidos. *Programa Nacional de Desarrollo Urbano y Vivienda, 1984–1988*. Mexico City: Elzevir Editores, August 1984. 156 pp.
——. *Programa Nacional de Ecologia, 1984-1988*. Mexico City: Elzevir Editores, August 1984. 271 pp.
Secretaria de Desarrollo Urbano y Ecologia (SEDUE). *Catalogo de Estudios y Proyectos, 1977–1982*. Three vols. Mexico City: mimeo, no date. 225, 160 and 281 pp.
——. *Manual de Fotointerpretación Urbana*. Mexico City: SEDUE, November 1984. 128 pp.
——. *Sistema de Información y Seguimiento de la Dirección General de Desarrollo Urbano*. Mexico City: mimeo, no date. 48 pp.
Tomlin, C. Dana. *The Map Analysis Package*. Draft version prepared as part of a doctoral dissertation in the Graduate School of Yale University. New Haven: mimeo, January 1980. 81 pp.

Case 20: Computers and development (Panama)

"Bajo el Agua del Canal". Videotape. Panama: Channel 11, 1983.

La Comision Mixta sobre el Ambiente Natural. *Plan de Manejo Integral de la Cuenca Hidrográfica del Canal de Panamá.* Translated into English (with additions) as "Proposals for a Plan of Comprehensive Management of the Panama Canal Watershed". Adopted by the Commission by Resolution No. 5 of June 30, 1983. Panama: mimeo, 1983. 15 pp.

"El Medio Ambiente". VHS Videotape. Panama: Channel 4, 1984.

Panama, Republic of. Asamblea Legislativa. *Proyecto del Ley No. 21: por el cual se Crean el Consejo Consultativa y la Dirección General del Area Canalera, y se Dictan Otras Medidas.* Draft version. Panama: mimeo, no date. 11 pp.

——. Ministerio de Desarrollo Agropecuario. Dirección de Recursos Nacionales Renovables (RENARE). *Situación Socio-Económica en la Cuenca Hidrográfica del Canal de Panamá.* Chapter 4. Draft. Panama: mimeo, no date. 101 pp.

——. Ministerio de Planificación y Política Económica. *National Development and Recovery of the Panama Canal Zone.* Working Document. Translated by the Panama Canal Company. Panama: mimeo, June 1978. 143 pp.

——. *Plan General de Usos del Suelo para el Area y la Cuenca Hidrográfica del Canal de Panamá.* 2nd ed. Panama: mimeo, September 1984. 143 pp. approx.

——. *Plan Nacional de Desarrollo, 1976–80.* Preliminary Version. Two vols. Panama: mimeo, 1976. 541 pp.

——. Ministerio de Vivienda. Dirección-General de Desarrollo Urbano. *Normas Técnicas para la Cuenca Hidrográfica del Lago Alajuela.* Panama: mimeo, November 1978. 57 pp.

——. *Proyecto de Ley No. 21 de 1985: Por el cual se crean el Consejo Consultativo y la Dirección General del Area Canalera, y se Dictan Otras Medidas.* Panama: mimeo, 1985. 11 pp.

Panama Canal Commission. *A Report on the Panama Canal Rain Forest.* By Frank H. Robinson. Panama: mimeo, January 1985. 70 pp.

Ricardo Riba, Jorge, ed. *Alternativas de Desarrollo y Uso de las Areas Revertidas: Informe del Seminario.* Panama: mimeo, July 1984. Approx. 150 pp.

Robert R. Nathan Associates Inc. and The Urban Institute. *Urban Development Assessment: Panama.* Two vols. Draft. Prepared for the United States Agency for International Development (USAID). Washington DC: mimeo, October 1984. Approx. 230 pp.

For Product Safety Concerns and Information please contact our EU representative GPSR@taylorandfrancis.com
Taylor & Francis Verlag GmbH, Kaufingerstraße 24, 80331 München, Germany

www.ingramcontent.com/pod-product-compliance
Ingram Content Group UK Ltd.
Pitfield, Milton Keynes, MK11 3LW, UK
UKHW021443080625
459435UK00011B/351

* 9 7 8 0 3 6 7 7 7 1 9 6 6 *